Algebra 2

LARS[illegible]
BOSW[illegible]
KANOLD
STIFF

Applications • Equations • Graphs

Chapter 7 Resource Book

The Resource Book contains the wide variety of blackline masters available for Chapter 7. The blacklines are organized by lesson. Included are support materials for the teacher as well as practice, activities, applications, and assessment resources.

McDougal Littell
A HOUGHTON MIFFLIN COMPANY
Evanston, Illinois • Boston • Dallas

Contributing Authors

The authors wish to thank the following individuals for their contributions to the Chapter 7 Resource Book.

Rose Elaine Carbone
José Castro
John Graham
Fr. Chris M. Hamlett
Edward H. Kuhar
Cheryl A. Leech
Ann C. Nagosky
Karen Ostaffe
Leslie Palmer
Ann Larson Quinn, Ph. D.
Chris Thibaudeau

ISBN: 0-618-02015-2

23456789-PBO- 04 03 02 01

Contents

7 *Powers, Roots, and Radicals*

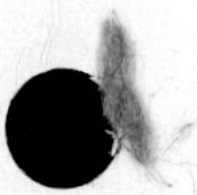

Contents

Contents

Descriptions of Resources

This Chapter Resource Book is organized by lessons within the chapter in order to make your planning easier. The following materials are provided:

Tips for New Teachers These teaching notes provide both new and experienced teachers with useful teaching tips for each lesson, including tips about common errors and inclusion.

Parent Guide for Student Success This guide helps parents contribute to student success by providing an overview of the chapter along with questions and activities for parents and students to work on together.

Prerequisite Skills Review Worked-out examples are provided to review the prerequisite skills highlighted on the Study Guide page at the beginning of the chapter. Additional practice is included with each worked-out example.

Strategies for Reading Mathematics The first page teaches reading strategies to be applied to the current chapter and to later chapters. The second page is a visual glossary of key vocabulary.

Lesson Plans and Lesson Plans for Block Scheduling This planning template helps teachers select the materials they will use to teach each lesson from among the variety of materials available for the lesson. The block-scheduling version provides additional information about pacing.

Warm-Up Exercises and Daily Homework Quiz The warm-ups cover prerequisite skills that help prepare students for a given lesson. The quiz assesses students on the content of the previous lesson. (Transparencies also available)

Activity Support Masters These blackline masters make it easier for students to record their work on selected activities in the Student Edition.

Alternative Lesson Openers An engaging alternative for starting each lesson is provided from among these four types: ***Application, Activity, Graphing Calculator,*** or ***Visual Approach.*** (Color transparencies also available)

Graphing Calculator Activities with Keystrokes Keystrokes for four models of calculators are provided for each Technology Activity in the Student Edition, along with alternative Graphing Calculator Activities to begin selected lessons.

Practice A, B, and C These exercises offer additional practice for the material in each lesson, including application problems. There are three levels of practice for each lesson: A (basic), B (average), and C (advanced).

Contents

Reteaching with Practice These two pages provide additional instruction, worked-out examples, and practice exercises covering the key concepts and vocabulary in each lesson.

Quick Catch-Up for Absent Students This handy form makes it easy for teachers to let students who have been absent know what to do for homework and which activities or examples were covered in class.

Cooperative Learning Activities These enrichment activities apply the math taught in the lesson in an interesting way that lends itself to group work.

Interdisciplinary Applications/Real-Life Applications Students apply the mathematics covered in each lesson to solve an interesting interdisciplinary or real-life problem.

Math and History Applications This worksheet expands upon the Math and History feature in the Student Edition.

Challenge: Skills and Applications Teachers can use these exercises to enrich or extend each lesson.

Quizzes The quizzes can be used to assess student progress on two or three lessons.

Chapter Review Games and Activities This worksheet offers fun practice at the end of the chapter and provides an alternative way to review the chapter content in preparation for the Chapter Test.

Chapter Tests A, B, and C These are tests that cover the most important skills taught in the chapter. There are three levels of test: A (basic), B (average), and C (advanced).

SAT/ACT Chapter Test This test also covers the most important skills taught in the chapter, but questions are in multiple-choice and quantitative-comparison format. (See *Alternative Assessment* for multi-step problems.)

Alternative Assessment with Rubrics and Math Journal A journal exercise has students write about the mathematics in the chapter. A multi-step problem has students apply a variety of skills from the chapter and explain their reasoning. Solutions and a 4-point rubric are included.

Project with Rubric The project allows students to delve more deeply into a problem that applies the mathematics of the chapter. Teacher's notes and a 4-point rubric are included.

Cumulative Review These practice pages help students maintain skills from the current chapter and preceding chapters.

CHAPTER 7

Tips for New Teachers

For use with Chapter 7

Chapter Support

Lesson 7.1

TEACHING TIP When students evaluate expressions where there is a rational exponent of the form $\frac{m}{n}$, ask them to first take the nth root and then raise the result to the mth power. This will keep their numbers smaller, allowing them to find the answer without using a calculator. For example, students are more likely to be able to evaluate

$$32^{2/5} = (32^{1/5})^2 = 2^2 = 4$$

than

$$32^{2/5} = (32^2)^{1/5} = 1024^{1/5} = 4.$$

TEACHING TIP Some scientific calculators do not calculate powers when the base is a negative number—instead they display an *error* message. Students who work with these calculators might think that it is impossible to take *any* root of a negative number. Remind students that it is not possible to take an even root of a negative number, but they can always find odd roots for any number.

COMMON ERROR Some students are confused by negative exponents and rational exponents with a numerator equal to 1. They might find a root when they should find the reciprocal and vice versa. Before completing exercises with *negative rational exponents*, make sure that students understand the difference between negative and rational exponents.

Lesson 7.2

TEACHING TIP Students can easily simplify a variable radical expression with an nth root, dividing each exponent in the radicand by n. When they do this, the *quotient* is the exponent of the variable *outside* the radical and the *remainder*—if it exists—is the exponent of the variable left *inside* the radical. For example, for $\sqrt[5]{a^5b^{13}}$, since $5 \div 5 = 1$ with no remainder, we get a^1 outside the radical. Since $13 \div 5 = 2$ with remainder 3, we get b^2 outside the radical and b^3 inside it. The final answer is $ab^2\sqrt[5]{b^3}$.

Lesson 7.3

TEACHING TIP A function can be considered as a *machine* that will produce an output for a given input:

x (input) → Function f → $f(x)$ (output)

This analogy can help students to understand *composition of functions*. In this manner, $f(g(x))$ is the result of using two machines, one immediately after the other:

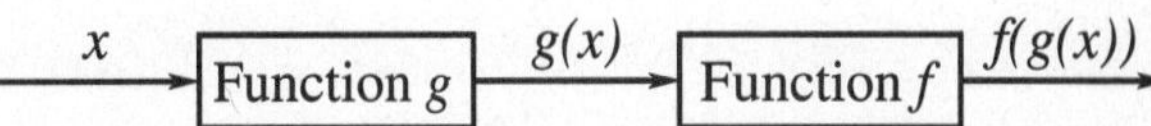

Students can first learn to evaluate composition of functions by evaluating $g(x)$ and plugging in that value into function $f(x)$. Once students are comfortable with this idea, introduce the new function $f(g(x))$ that will yield the same results as before but *directly:*

x → Function $f(g)$ → $f(g(x))$

This model might also be helpful to show the difference between $f(g(x))$ and $g(f(x))$ as well as to discuss the domain of $f(g(x))$ based on the domains of $f(x)$ and $g(x)$.

Lesson 7.4

TEACHING TIP Use again the model of functions as *machines* to help students understand the idea of inverse functions:

x → Function f → $f(x)$ → Function f^{-1} → x

and

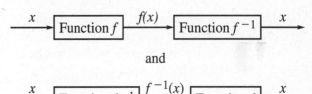

x → Function f^{-1} → $f^{-1}(x)$ → Function f → x

TEACHING TIP After this lesson you can define a *one-to-one function* as a function whose inverse is also a function. The graph of a one-to-one function must pass both the *vertical* and *horizontal line tests*. A one-to-one function has *exactly* one output for each possible input.

COMMON ERROR Many students do not understand the need to show that $f(f^{-1}(x)) = x$ *and* $f^{-1}(f(x)) = x$ to verify that two functions are inverses of each other. Use the following example to show them that they must check *both* conditions. Take $f(x) = x^2$ and $g(x) = \sqrt{x}$. We have

Tips for New Teachers

For use with Chapter 7

Lesson 7.4 (Cont.)

$$f(g(x)) = (\sqrt{x})^2 = (x^{1/2})^2 = x.$$

However, from Lesson 7.2 we know that

$$g(f(x)) = \sqrt{(x^2)} = |x|.$$

This is a good example because the domains for $f(g(x))$ and $g(f(x))$ are not the same, something you might want to discuss with your students.

Lesson 7.5

TEACHING TIP The technique used in this lesson to graph square and cube root functions can also be used for any other family of functions. In general, the graph of $g(x)$, where $g(x) = f(x - h) + k$, is just the graph of $f(x)$ shifted h units to the right and k units up. You can show students the connection between this technique and the graphs for quadratic and absolute value functions, such as $y = a(x - h)^2 + k$ and $y = a|x - h| + k$.

TEACHING TIP After completing several examples, ask your students what the domain and range are for *all* cube root functions—both are all real numbers. Students can find the domain and range of square root functions without graphing the function. For the domain, they just need to solve an inequality where the radicand is greater than or equal to zero. As for the range, it will always be either $y \geq k$ if the radical is positive or $y \leq k$ if the radical is negative.

Lesson 7.6

COMMON ERROR When solving equations with one or two radicals, some students raise both sides of the equation to the same power *before* they solve for the radical. For example, they might do things such as:

$$\sqrt{4x - 7} + 2 = 5$$
$$(\sqrt{4x - 7} + 2)^2 = (5)^2$$
$$4x - 7 + 4 = 25$$

Show students their error by asking them to use FOIL on the left side of the equation. The radical will not go away. This will show students that to make the radical disappear they need to isolate it on one side of the equation before taking the powers.

Lesson 7.7

TEACHING TIP The mean and the median are close in value many times, and some students might not see the value of knowing both of them. Use a set of data such as 1, 2, 3, 4, 1000 to show students that it might be necessary to know both measures of central tendency to understand the data. For the previous data the mean is 202 and the median is 3. Ask your students to suppose they do not know the data but they are given the mean and the median. Which of the two measures of central tendency is more representative of the data? Why? Would it help them to understand the data if they knew the mode? What about knowing any of the measures of dispersion?

Outside Resources

BOOKS/PERIODICALS

Naraine, Bishnu. "An Alternative Approach to Solving Radical Equations." *Mathematics Teacher* (March 1993), pp. 204–205.

ACTIVITIES/MANIPULATIVES

Coes, Loring, III. "More Functions of a Toy Balloon." *Mathematics Teacher* (April 1997), pp. 290–294, 300–302.

Alge-Solids. 3-D models to demonstrate powers of x and y. White Plains NY. Dale Seymour Publications.

SOFTWARE

Cabri Geometry II., Create figures to experiment and explore algebraic principals. Vernon Hills, IL. ETA.

NAME _______________ DATE _______

Parent Guide for Student Success

For use with Chapter 7

Chapter Overview One way that you can help your student succeed in Chapter 7 is by discussing the lesson goals in the chart below. When a lesson is completed, ask your student to interpret the lesson goals for you and to explain how the mathematics of the lesson relates to one of the key applications listed in the chart.

Lesson Title	Lesson Goals	Key Applications
7.1: *n*th Roots and Rational Exponents	Evaluate *n*th roots of real numbers using both radical and rational exponent notation. Use *n*th roots to solve real-life problems.	• Science • Spillway of a Dam • Inflation
7.2: Properties of Rational Exponents	Use properties of rational exponents to evaluate and simplify expressions and to solve real-life problems.	• Biology • Pinhole Camera • Music
7.3: Power Functions and Function Operations	Perform operations with functions and use power functions and function operations to solve real-life problems.	• Business • Coat Sale • Paleontology
7.4: Inverse Functions	Find inverses of linear and nonlinear functions.	• Astronomy • Exchange Rate • Bowling
7.5: Graphing Square Root and Cube Root Functions	Graph square root and cube root functions and use square root and cube root functions to find real-life quantities.	• Amusement Parks • Racing • Storms at Sea
7.6: Solving Radical Equations	Solve equations that contain radicals or rational exponents and use radical equations to solve real-life problems.	• Beaufort Wind Scale • Science: Dinosaurs • America's Cup
7.7: Statistics and Statistical Graphs	Use measures of central tendency and measures of dispersion to describe data sets. Use box-and-whisker plots and histograms to represent data graphically.	• Women's Basketball • Scratch Protection • Political Survey

Test-Taking Strategy

Always **check your solution** against the original problem. When possible **use a different method** to check so you don't repeat an error. Ask your student why going back to the original problem is especially important in Lesson 7.6. Have your student show you an example from the chapter of how to check using different steps than those used to solve the problem.

NAME ______________________________ DATE ____________

Parent Guide for Student Success

For use with Chapter 7

Key Ideas **Your student can demonstrate understanding of key concepts by working through the following exercises with you.**

Lesson	Exercise
7.1	The relationship between a cubic foot F and x cubic meters can be approximated by the equation $F = 0.0283x^3$. A room has a volume of 1615 cubic feet. How many cubic meters is that?
7.2	Simplify the expression $(y \cdot y^{1/3})^{3/5}$. Assume all variables are positive.
7.3	The sale price of a pair of shoes is defined by the function $f(x) = 0.8x$ and the sales tax on a purchase is defined by the function $g(x) = 1.04x$. Find $g(f(x))$. Use $g(f(x))$ to find the sale price with tax of a \$65 pair of shoes.
7.4	Find the inverse function of $f(x) = 3 - \frac{1}{4}x^2, x \geq 0$.
7.5	The maximum walking speed S (in feet per second) of an animal can be modeled by $S = \sqrt{32L}$, where L is the length of the animal's legs in feet. Use a graph to determine the length of the legs of an early mammal that had a maximum walking speed of 6 feet per second.
7.6	Solve the equation $-\sqrt[3]{(4 + 5x)} = \sqrt[3]{(8 - 6x)}$. Check for extraneous solutions.
7.7	The Valley High School football team scored the following points in the first five games of the season: 13, 7, 10, 21, 3. Find the mean and standard deviation of the scores.

Home Involvement Activity

Directions: Record how many hours each family member spends watching television each day for a few days. Find the mean, median, mode, range, and standard deviation of the hours. Interpret the results in terms of your family's viewing habits.

Answers

7.1: about 38.5 m^3 **7.2:** $y^{4/5}$ **7.3:** $g(f(x)) = 0.832x$; \$54.08 **7.4:** $f^{-1}(x) = \pm 2\sqrt{3 - x}$
7.5: 1.125 ft **7.6:** 12 **7.7:** 10.8 points; about 6.08 points

CHAPTER 7

NAME ______________________ DATE ________

Prerequisite Skills Review

For use before Chapter 7

Chapter Support

EXAMPLE 1

Rewriting an Equation with More Than One Variable

Solve the equation for y.

$5x + \frac{2}{5}y = -3$

SOLUTION

$5x + \frac{2}{5}y = -3$	Write original equation.
$\frac{2}{5}y = -5x - 3$	Subtract $5x$ from each side.
$y = \frac{-25x - 15}{2}$	Multiply each side by $\frac{5}{2}$.

Exercises for Example 1

Solve the equation for *y*.

1. $-4x + 3y = 5$ **2.** $x - \frac{1}{6}y = -2$ **3.** $2x = -y + 8$

EXAMPLE 2

Factoring a Trinomial

Factor the trinomial.

a. $x^2 + 3x - 18$ **b.** $3x^2 - 16x + 21$

SOLUTION

a. You want $x^2 + 3x - 18 = (x + m)(x + n)$ where $mn = -18$ and $m + n = 3$.

Factors of -18 *(mn)*	$-1, 18$	$1, -18$	$-2, 9$	$2, -9$	$-3, 6$	$3, -6$
Sum of factors $(m + n)$	17	-17	7	-7	3	-3

The table shows that $m = -3$ and $n = 6$. So, $x^2 + 3x - 18 = (x - 3)(x + 6)$.

b. You want $3x^2 - 16x + 21 = (kx + m)(lx + n)$ where k and l are factors of 3 and m and n are (negative) factors of 21. Check possible factorizations by multiplying.

$(3x - 21)(x - 1) = 3x^2 - 24x + 21$ $(3x - 1)(x - 21) = 3x^2 - 64x + 21$

$(3x - 3)(x - 7) = 3x^2 - 24x + 21$ $(3x - 7)(x - 3) = 3x^2 - 16x + 21$ ✓

The correct factorization is $3x^2 - 16x + 21 = (3x - 7)(x - 3)$.

Exercises for Example 2

Factor the trinomial.

4. $x^2 - 11x + 28$ **5.** $2x^2 + 7x - 15$ **6.** $x^2 - 5x - 24$

NAME ______________________ DATE ________

Prerequisite Skills Review

For use before Chapter 7

EXAMPLE 3 Simplifying Algebraic Expressions

Simplify the expression.

a. $(x^2yz^3)^4$

b. $\dfrac{a^2b^{-3}}{a^{-4}b^5}$

SOLUTION

a. $(x^2yz^3)^4 = (x^2)^4y^4(z^3)^4$ Power of a product property

$= x^8y^4z^{12}$ Power of a power property

b. $\dfrac{a^2b^{-3}}{a^{-4}b^5} = a^{2-(-4)}b^{-3-5}$ Quotient of powers property

$= a^6b^{-8}$ Simplify exponents.

$= \dfrac{a^6}{b^8}$ Negative exponent property

Exercises for Example 3

Simplify the expression.

7. $(a^{-3}b^4c^{-1})^2$

8. $\left(\dfrac{x^{-2}}{y^3}\right)^3$

9. $(5x^{-3}y^2)^{-2}$

EXAMPLE 4 Adding, Subtracting, and Multiplying Polynomials

Perform the indicated operation.

$4x^3(x + 3)$

SOLUTION

$4x^3(x + 3) = 4x^3(x) + 4x^3(3)$ Distributive property

$= 4x^4 + 12x^3$ Simplify.

Exercises for Example 4

Perform the indicated operation.

10. $2x^4(x - 6)$

11. $(4y - 1)^2$

12. $(x^2 - 3x) - (5x - 2)$

NAME ______________________________ DATE __________

Strategies for Reading Mathematics

For use with Chapter 7

Strategy: Reading Statistical Graphs

You have learned how to graph functions and relations in the plane, in particular linear, quadratic, and other polynomial functions. Statistical graphs are quite different. Their purpose is to give a visual picture of a set of data.

Suppose the numbers of representatives that states have in the United States House of Representatives are: 7, 1, 6, 4, 52, 6, 6, 1, 23, 11, 2, 2, 20, 10, 5, 4, 6, 7, 2, 8, 10, 16, 8, 5, 9, 1, 3, 2, 2, 13, 3, 31, 12, 1, 19, 6, 5, 21, 2, 6, 1, 9, 30, 3, 1, 11, 9, 3, 9, 1. It is hard to get a picture of a large data set listed this way.

One type of visual picture of the data is given by a box-and-whisker plot. The plot below shows that the data is centered at 6 and that half the values lie between 2 and 10, inclusive.

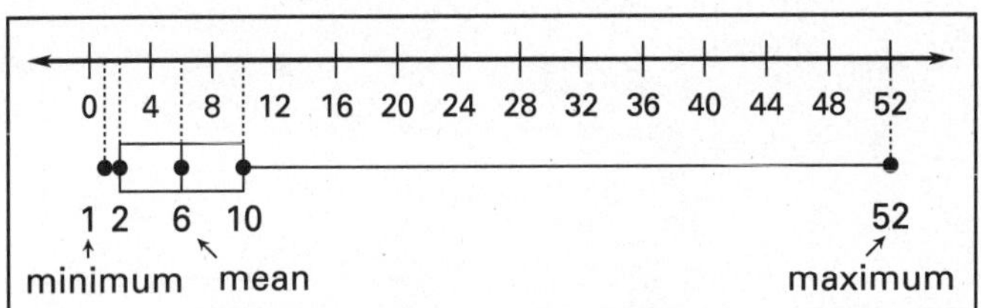

Another type of visual picture of the data is given by a histogram. A histogram for the data above is shown beside Questions 1–4.

> **STUDY TIP**
> ***Reading a Box-and-Whisker Plot***
> A box-and-whisker plot gives a quick visual representation of a set of data. From it you can find the smallest data value, the middle value and the largest value, as well as the interval that holds the middle 50% of the data values.

> **STUDY TIP**
> ***Reading a Histogram***
> A histogram shows the distribution of a data set. The horizontal axis is marked with equal intervals to cover the full range of the data. The vertical axis gives the number of data values within each interval.

Questions

1. What is the range of the data set given above? Which type of graph allows you to find the range easily?

2. What is the width of the interval containing the middle 50% of the data values?

3. Give the percent of data values that lie in each interval.

a. 1–5

b. 16–20

c. 36–40

4. How can you find the total number of data values from a histogram?

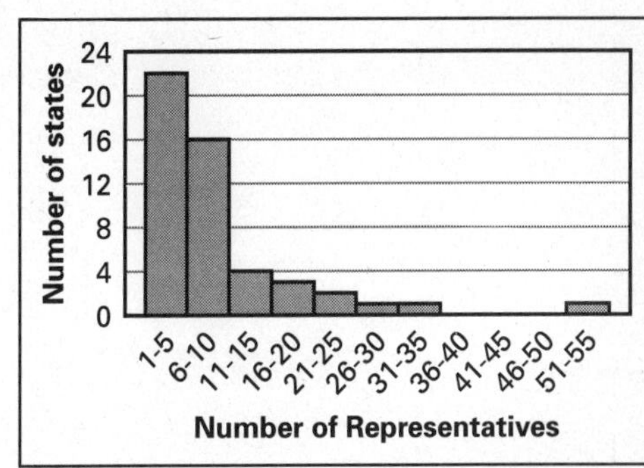

NAME ______________________________ DATE __________

Strategies for Reading Mathematics

For use with Chapter 7

Visual Glossary

The Study Guide on page 400 lists the key vocabulary for Chapter 7 as well as reviews vocabulary from previous chapters. Use the page references on page 400 or the Glossary in the textbook to review key terms from prior chapters. Use the visual glossary below to help you understand some of the key vocabulary in Chapter 7. You may want to copy these diagrams into your notebook and refer to them as you complete the chapter.

GLOSSARY

nth* root of *a (p. 401) For an integer n greater than 1, if $b^n = a$, then b is an nth root of a. Written as $\sqrt[n]{a}$ or $a^{1/n}$.

radical function (p. 431) A function that contains a radical, such as $y = \sqrt{x}$ or $y = \sqrt[3]{x}$.

inverse relation (p. 422) A relation that maps the output values of a relation back to their original input values.

Graphing a Radical Fraction

A function like $y = \sqrt{x}$ or $y = \sqrt[3]{x}$ is a radical function. You can graph a radical function by plotting points. Note any restrictions on the domain.

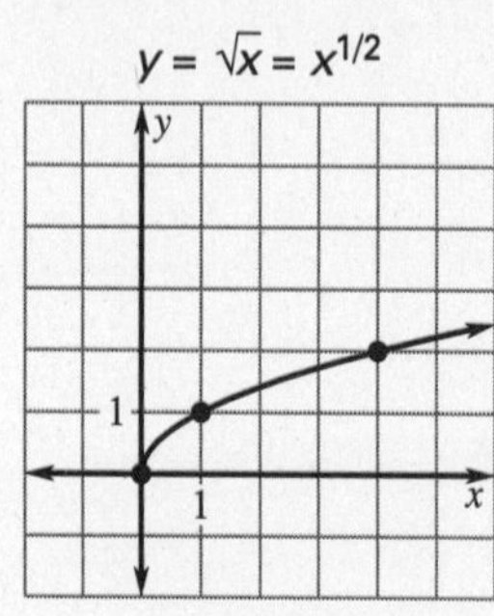

Domain: $x \geq 0$.
Even roots are only defined for positive values.

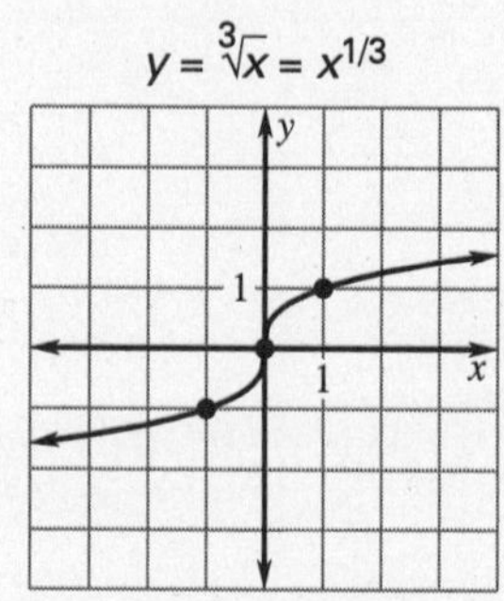

Domain: all real numbers.
Odd roots defined for all real numbers.

Finding the Inverse of a Function

To find the inverse of a function you switch the roles of x and y and solve for y:

$y = x^3$	Original function
$x = y^3$	Switch x and y.
$y = \sqrt[3]{x}$	Solve for y.

The graph of $f^{-1}(x)$ is the reflection of $f(x)$ in the line $y = x$.

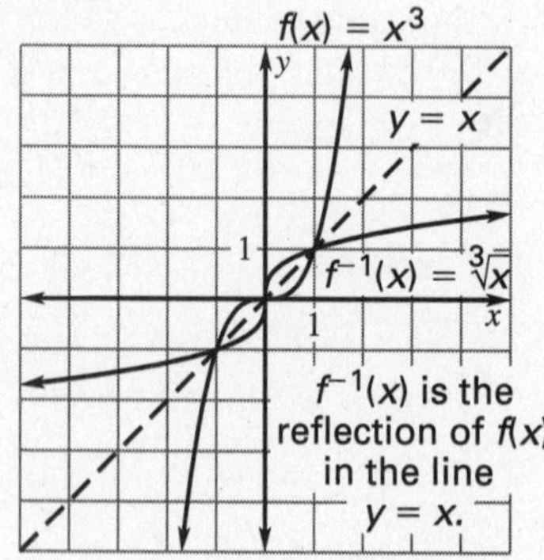

LESSON 7.1

TEACHER'S NAME ______________ CLASS ______ ROOM ______ DATE ______

Lesson Plan

1-day lesson (See *Pacing the Chapter,* TE pages 398C–398D)

For use with pages 401–406

1. **Evaluate *n*th roots of real numbers using both radical notation and rational exponent notation.**
2. **Use *n*th roots to solve real-life problems.**

State/Local Objectives ______________

____

✓ Check the items you wish to use for this lesson.

STARTING OPTIONS

____ Prerequisite Skills Review: CRB pages 5–6
____ Strategies for Reading Mathematics: CRB pages 7–8
____ Warm-Up or Daily Homework Quiz: TE pages 401 and 385, CRB page 11, or Transparencies

TEACHING OPTIONS

____ Motivating the Lesson: TE page 402
____ Lesson Opener (Application): CRB page 12 or Transparencies
____ Examples 1–6: SE pages 401–403
____ Extra Examples: TE pages 402–403 or Transparencies
____ Closure Question: TE page 403
____ Guided Practice Exercises: SE page 404

APPLY/HOMEWORK

Homework Assignment

____ Basic 13–22, 24–62 even, 68, 73–87 odd
____ Average 13–28, 30–64 even, 68, 73–87 odd
____ Advanced 13–28, 30–68 even, 69–71, 73–89 odd

Reteaching the Lesson

____ Practice Masters: CRB pages 13–15 (Level A, Level B, Level C)
____ Reteaching with Practice: CRB pages 16–17 or Practice Workbook with Examples
____ Personal Student Tutor

Extending the Lesson

____ Applications (Interdisciplinary): CRB page 19
____ Challenge: SE page 406; CRB page 20 or Internet

ASSESSMENT OPTIONS

____ Checkpoint Exercises: TE pages 402–403 or Transparencies
____ Daily Homework Quiz (7.1): TE page 406, CRB page 23, or Transparencies
____ Standardized Test Practice: SE page 406; TE page 406; STP Workbook; Transparencies

Notes ______________

____

____

LESSON 7.1

TEACHER'S NAME ______________________ CLASS ______ ROOM ______ DATE ______

Lesson Plan for Block Scheduling

Half-day lesson (See *Pacing the Chapter,* TE pages 398C–398D) For use with pages 401–406

GOALS

1. **Evaluate *n*th roots of real numbers using both radical notation and rational exponent notation.**
2. **Use *n*th roots to solve real-life problems.**

State/Local Objectives ______________________

____

____

CHAPTER PACING GUIDE	
Day	**Lesson**
1	Assess Ch. 6; **7.1(all)**
2	7.2 (all)
3	7.3 (all); 7.4(begin)
4	7.4 (end); 7.5(all)
5	7.6 (all)
6	7.7 (all)
7	Review/Assess Ch. 7

✓ Check the items you wish to use for this lesson.

STARTING OPTIONS

____ Prerequisite Skills Review: CRB pages 5–6
____ Strategies for Reading Mathematics: CRB pages 7–8
____ Warm-Up or Daily Homework Quiz: TE pages 401 and 385, CRB page 11, or Transparencies

TEACHING OPTIONS

____ Motivating the Lesson: TE page 402
____ Lesson Opener (Application): CRB page 12 or Transparencies
____ Examples 1–6: SE pages 401–403
____ Extra Examples: TE pages 402–403 or Transparencies
____ Closure Question: TE page 403
____ Guided Practice Exercises: SE page 404

APPLY/HOMEWORK

Homework Assignment

____ Block Schedule: 13–28, 30–64 even, 68, 73–87 odd

Reteaching the Lesson

____ Practice Masters: CRB pages 13–15 (Level A, Level B, Level C)
____ Reteaching with Practice: CRB pages 16–17 or Practice Workbook with Examples
____ Personal Student Tutor

Extending the Lesson

____ Applications (Interdisciplinary): CRB page 19
____ Challenge: SE page 406; CRB page 20 or Internet

ASSESSMENT OPTIONS

____ Checkpoint Exercises: TE pages 402–403 or Transparencies
____ Daily Homework Quiz (7.1): TE page 406, CRB page 23, or Transparencies
____ Standardized Test Practice: SE page 406; TE page 406; STP Workbook; Transparencies

Notes ______________________

____

____

LESSON 7.1

NAME ______________________ DATE __________

Available as a transparency

WARM-UP EXERCISES

For use before Lesson 7.1, pages 401–406

Evaluate the expression.

1. $\sqrt{9}$
2. $-\sqrt{121}$
3. $(\sqrt{25})^2$

Solve each equation.

4. $x^2 = 49$
5. $(x - 1)^2 = 64$

Lesson 7.1

DAILY HOMEWORK QUIZ

For use after Lesson 6.9, pages 379–386

1. Write a cubic function whose graph passes through $(-3, 0)$, $(-1, 0)$, $(1, -8)$, and $(4, 0)$.

2. Show that the second-order differences for $f(x) = 3x^2 - 4x + 2$ are nonzero and constant.

3. Use finite differences and a system of equations to find a polynomial function that fits the data.

x	1	2	3	4	5
$f(x)$	2	0	4	20	54

LESSON

7.1

NAME ____________________ DATE __________

Available as a transparency

Application Lesson Opener

For use with pages 401–406

As you know, a square with area A square units has sides of length x units, where $x = \sqrt{A}$.

Similarly, a cube with volume V cubic units has sides of length x units, where $x = \sqrt[3]{V}$. The number x is called the *cube* root of V, because $x^3 = V$.

For example, $2^3 = 8$, so 2 is the cube root of 8, or $2 = \sqrt[3]{8}$. If the volume of a cube is 8 cm^2, each edge of the cube is 2 cm long.

Find the side length of each square or cube.

1. Square with area 121 cm^2

2. Square with area 64 ft^2

3. Square with area 81 m^2

4. Square with area 25 cm^2

5. Square with area 1.44 in.^{2}

6. Cube with volume 27 m^3

7. Cube with volume 216 cm^3

8. Cube with volume 125 in.^{3}

9. Cube with volume 343 mm^3

10. Cube with volume 0.512 ft^3

LESSON 7.1

NAME ______________________ DATE __________

Practice A

For use with pages 401–406

Rewrite the expression using rational exponent notation.

1. $\sqrt[3]{11}$
2. $\sqrt[4]{5}$
3. $\sqrt[5]{23}$
4. $\sqrt{7}$
5. $\sqrt[3]{17}$
6. $\sqrt[6]{2}$
7. $\sqrt[4]{8}$
8. $\sqrt[3]{15}$
9. $\sqrt{10}$
10. $\sqrt[7]{3}$
11. $\sqrt[5]{6}$
12. $\sqrt[8]{21}$

Rewrite the expression using radical notation.

13. $2^{1/3}$
14. $5^{1/4}$
15. $11^{1/2}$
16. $6^{1/5}$
17. $23^{1/7}$
18. $31^{1/4}$
19. $103^{1/2}$
20. $17^{1/3}$
21. $4^{1/3}$
22. $7^{1/8}$
23. $8^{1/5}$
24. $12^{1/14}$

Evaluate the expression without using a calculator.

25. $\sqrt[3]{8}$
26. $\sqrt[4]{81}$
27. $\sqrt[5]{32}$
28. $\sqrt[3]{64}$
29. $\sqrt[4]{1}$
30. $\sqrt[3]{125}$
31. $(1)^{1/6}$
32. $(16)^{1/4}$
33. $(-8)^{1/3}$

Evaluate the expression using a calculator. Round the result to two decimal places.

34. $\sqrt[3]{5}$
35. $\sqrt[3]{24}$
36. $\sqrt[4]{10}$
37. $\sqrt[4]{3}$
38. $\sqrt[5]{16}$
39. $\sqrt[5]{8}$
40. $(6)^{1/5}$
41. $(12)^{1/3}$
42. $(7)^{1/4}$
43. $(4)^{1/5}$
44. $(29)^{1/3}$
45. $(126)^{1/6}$

46. ***Geometry*** Find the length of an edge of the cube shown below.

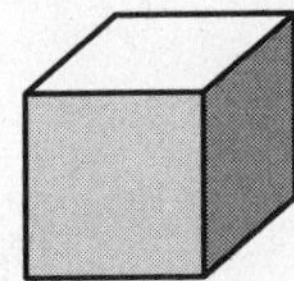

Volume = 216 in.^{3}

47. ***Geometry*** Find the length of an edge of the cube shown below.

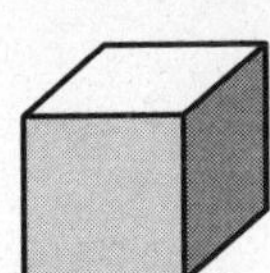

Volume = 527 cm^3

Lesson 7.1

LESSON

7.1

NAME ______________________ DATE __________

Practice B

For use with pages 401–406

Rewrite the expression using rational exponent notation.

1. $\sqrt[3]{7}$
2. $(\sqrt[3]{5})^2$
3. $(\sqrt{11})^5$
4. $(\sqrt[6]{12})^{10}$
5. $(\sqrt[3]{15})^7$
6. $(\sqrt[3]{-9})^5$
7. $(\sqrt[7]{-42})^2$
8. $(\sqrt[3]{-10})^8$

Rewrite the expression using radical notation.

9. $19^{1/3}$
10. $43^{1/5}$
11. $6^{2/3}$
12. $9^{4/3}$
13. $8^{3/4}$
14. $(-6)^{2/3}$
15. $(-10)^{4/3}$
16. $(-14)^{3/7}$

Evaluate the expression without using a calculator.

17. $8^{4/3}$
18. $36^{3/2}$
19. $16^{3/4}$
20. $81^{3/2}$
21. $64^{2/3}$
22. $32^{2/5}$
23. $4^{5/2}$
24. $(-64)^{1/3}$
25. $(-8)^{5/3}$

Evaluate the expression using a calculator. Round the result to two decimal places.

26. $\sqrt[4]{49}$
27. $\sqrt[9]{19{,}422}$
28. $\sqrt[5]{-122}$
29. $(215)^{1/5}$
30. $(-15)^{1/3}$
31. $(116)^{1/6}$
32. $(132)^{2/3}$
33. $(28)^{5/2}$
34. $(-112)^{8/3}$
35. ***Geometry*** Find the radius of a sphere with a volume of 382 cubic centimeters.

Solve the equation. Round your answer to two decimal places when appropriate.

36. $x^2 - 5 = 139$
37. $5x^3 = 3650$
38. $(x - 7)^3 = 729$

Water and Ice **In Exercises 39–42, use the following information.**

Water, in its liquid state, has a density of 0.9971 grams per cubic centimeter. Ice has a density of 0.9168 grams per cubic centimeter. You fill a cubical container with 510 grams of liquid water. A different cubical container is filled with 510 grams of solid water (ice).

39. Find the volume of the container filled with liquid water.
40. Find the length of the edges of the container in Exercise 39.
41. Find the volume of the container filled with ice.
42. Find the length of the edges of the container in Exercise 41.

NAME ______________________ DATE ____________

Reteaching with Practice

For use with pages 401–406

EXAMPLE 3 Approximating a Root with a Calculator

Use a graphing calculator to approximate $(\sqrt[3]{-7})^2$.

SOLUTION

Begin by rewriting $(\sqrt[3]{-7})^2$ as $(-7)^{2/3}$. Then enter the following:

Keystrokes: ((-) 7) ^ (2 ÷ 3) ENTER **Display:** 3.65930571

$(\sqrt[3]{-7})^2 \approx 3.66$

Notice that the negative radicand of -7 was enclosed in parentheses. If parentheses were not used, only the 7 would have been raised to the two-thirds power, and the result would have been negative.

Exercises for Example 3

Evaluate the expressions using a calculator. Round the result to two decimal places.

13. $\sqrt[4]{252}$ **14.** $\sqrt[3]{-2111}$ **15.** $(\sqrt[3]{56})^4$

16. $(\sqrt[3]{-140})^6$ **17.** $\sqrt[8]{25{,}102}$ **18.** $(\sqrt[3]{5})^3$

EXAMPLE 4 Solving Equations Using nth Roots

Solve the equation.

a. $-5x^2 = -30$

$x^2 = 6$ Divide each side by -5.

$x = \pm\sqrt{6}$ Take square root of each side.

$x \approx \pm 2.45$ Round result.

b. $(x + 4)^3 = 27$

$x + 4 = \sqrt[3]{27}$ Take cube root of each side.

$x + 4 = 3$ Simplify.

$x = -1$ Subtract 4 from each side.

Exercises for Example 4

Solve the equation. Round your answer to two decimal places when appropriate.

19. $x^4 = 87$ **20.** $2x^3 = 92$ **21.** $(x - 1)^5 = 12$

LESSON 7.1

NAME ______________________________ DATE ________

Quick Catch-Up for Absent Students

For use with pages 401–406

The items checked below were covered in class on (date missed) __________

Lesson 7.1: *n*th Roots and Rational Exponents

____ **Goal 1:** Evaluate *n*th roots of real numbers using both radical notation and rational exponent notation. (pp. 401, 402)

Material Covered:

____ Example 1: Finding *n*th Roots

____ Example 2: Evaluating Expressions with Rational Exponents

____ Student Help: Study Tip

____ Example 3: Approximating a Root with a Calculator

____ Example 4: Solving Equations Using *n*th Roots

Vocabulary:

*n*th root of *a*, p. 401 index, p. 401

____ **Goal 2:** Use *n*th roots to solve real-life problems. (p. 403)

Material Covered:

____ Example 5: Evaluating a Model with *n*th Roots

____ Example 6: Solving an Equation Using an *n*th Root

Vocabulary:

____ Other (specify) ______________________________

____

Homework and Additional Learning Support

____ Textbook (specify) pp. 404–406

____

____ *Reteaching with Practice* worksheet (specify exercises) __________

____ *Personal Student Tutor* for Lesson 7.1

NAME ______________________ DATE __________

Interdisciplinary Application

For use with pages 401–406

Johannes Kepler

SCIENCE Johannes Kepler (1571–1630) was a German astronomer and mathematician. He was a professor of mathematics and had many discoveries in mathematics including, identifying two new regular polyhedra and writing the first proof for logarithms.

Johannes Kepler is famous for his three laws of planetary motion which determine the path a planet travels around the sun. The first law states that each planet orbits the sun in an elliptical path with the sun at one focus. The second law explains how the speed of a planet increases as its distance from the sun decreases. The third law establishes a relationship between the average distance of the planet from the sun and the time to complete one revolution around the sun. Using his three laws of planetary motion, Kepler constructed the most exact astronomical tables.

In Exercises 1–3, use the following information.

Kepler's third law of planetary motion states the period P (in years where one year is 365.25 days) of each planet in our solar system is related to the planet's mean distance a (in astronomical units) from the sun. This law can be modeled by $P^2 = a^3$.

1. Given the period of each planet, find the mean distance from the sun. Copy and complete the table.

Planet	*P*	*a*
Mercury	0.241	
Venus	0.615	
Earth	1.000	1.000
Mars	1.881	
Jupiter	11.861	
Saturn	29.457	
Uranus	84.008	
Neptune	164.784	
Pluto	248.350	

2. Find the mean distance of the asteroid belt if the period is roughly between 2.828 and 6.547 years.
3. Between which two planets does the asteroid belt lie?

NAME ______________________ DATE __________

Challenge: Skills and Applications

For use with pages 401–406

In Exercises 1–3, rationalize each denominator, and express the fraction in simplest form.

Example $\frac{1}{\sqrt[5]{9}}$

Solution $\frac{1}{\sqrt[5]{9}} = \frac{1}{(3^2)^{1/5}} = \frac{1}{3^{2/5}} = \frac{1}{3^{2/5}} \cdot \frac{3^{3/5}}{3^{3/5}} = \frac{3^{3/5}}{3} = \frac{\sqrt[5]{27}}{3}$

1. $\frac{1}{\sqrt[4]{5}}$

2. $\frac{2}{\sqrt[3]{16}}$

3. $\frac{9}{\sqrt[5]{81}}$

4. a. Use your calculator to evaluate 7^x for $x = \frac{1}{3}, \frac{1}{4}, \frac{1}{10}, \frac{1}{50}, \frac{1}{100}$.

b. What seems to be true of the value of 7^x as $x \to 0$?

c. Make a conjecture about the values of $x^{1/x}$ as $x \to \infty$. Use your calculator to evaluate this expression for $x = 5, 10, 20, 50, 100$, and 1000. Was your conjecture correct?

d. Make a conjecture about x^x as $x \to 0$. Use your calculator to evaluate this expression for $x = \frac{1}{5}, \frac{1}{10}, \frac{1}{20}, \frac{1}{50}, \frac{1}{100}$, and $\frac{1}{1000}$.

In Exercises 5–10, solve each question. Express your answer(s) in simplest radical form. (*Hint:* $(x^{2/3})^7 = x$)

5. $(x - 1)^{4/3} = 16$

6. $x^{1/2} = 2x^{1/3}$

7. $x^{1/2} = 3x^{-4/3}$

8. $x^{2/5} - 3x^{1/5} + 2 = 0$
(*Hint:* Let $w = x^{1/5}$.)

9. $125x^{-3/2} = 8$

10. $x^{-2/3} - 2x^{-1/3} - 3 = 0$

11. Use rational exponents to prove that for any positive number x and any integers m and n, $\sqrt[m]{x^n} = \left(\sqrt[m]{x}\right)^n$.

LESSON 7.2

TEACHER'S NAME ______________________ CLASS ______ ROOM ______ DATE ______

Lesson Plan

2-day lesson (See *Pacing the Chapter,* TE pages 398C–398D) For use with pages 407–414

1. Use properties of rational exponents to evaluate and simplify expressions.
2. Use properties of rational exponents to solve real-life problems.

State/Local Objectives ____

____

✓ Check the items you wish to use for this lesson.

STARTING OPTIONS

____ Homework Check: TE page 404; Answer Transparencies
____ Warm-Up or Daily Homework Quiz: TE pages 407 and 406, CRB page 23, or Transparencies

TEACHING OPTIONS

____ Lesson Opener (Activity): CRB page 24 or Transparencies
____ Examples: Day 1: 1–5, SE pages 407–409; Day 2: 6–9, SE pages 409–410
____ Extra Examples: Day 1: TE pages 408–409 or Transp.; Day 2: TE pages 409–410 or Transp.; Internet
____ Closure Question: TE page 410
____ Guided Practice: SE page 411 Day 1: Exs. 1–12; Day 2: Exs. 13–21

APPLY/HOMEWORK

Homework Assignment

____ Basic Day 1: 22–48 even, 50–52, 55–63, 69, 71, 77; Day 2: 33, 41, 49, 53–55, 64–67, 68–90 even, 98, 101–111 odd; Quiz 1: 1–22
____ Average Day 1: 22–50 even, 56–84 even, 90–91; Day 2: 33, 41, 49, 55, 67, 81, 89, 92–94, 97, 101–111 odd; Quiz 1: 1–22
____ Advanced Day 1: 22–80 even, 81–88, 90–92; Day 2: 33, 41, 49, 55, 67, 81, 89, 93–99, 101–111 odd; Quiz 1: 1–22

Reteaching the Lesson

____ Practice Masters: CRB pages 25–27 (Level A, Level B, Level C)
____ Reteaching with Practice: CRB pages 28–29 or Practice Workbook with Examples
____ Personal Student Tutor

Extending the Lesson

____ Applications (Real-Life): CRB page 31
____ Challenge: SE page 413; CRB page 32 or Internet

ASSESSMENT OPTIONS

____ Checkpoint Exercises: Day 1: TE pages 408–409 or Transp.; Day 2: TE pages 409–410 or Transp.
____ Daily Homework Quiz (7.2): TE page 414, CRB page 36, or Transparencies
____ Standardized Test Practice: SE page 413; TE page 414; STP Workbook; Transparencies
____ Quiz (7.1–7.2): SE page 414; CRB page 33

Notes ____

____

____

Lesson 7.2

LESSON 7.2

TEACHER'S NAME ______________________ CLASS ______ ROOM ______ DATE ______

Lesson Plan for Block Scheduling

1-day lesson (See *Pacing the Chapter,* TE pages 398C–398D) For use with pages 407–414

GOALS
1. Use properties of rational exponents to evaluate and simplify expressions.
2. Use properties of rational exponents to solve real-life problems.

State/Local Objectives ______________________________

CHAPTER PACING GUIDE	
Day	**Lesson**
1	Assess Ch. 6; 7.1(all)
2	**7.2 (all)**
3	7.3 (all); 7.4(begin)
4	7.4 (end); 7.5(all)
5	7.6 (all)
6	7.7 (all)
7	Review/Assess Ch. 7

✓ Check the items you wish to use for this lesson.

STARTING OPTIONS

____ Homework Check: TE page 404; Answer Transparencies
____ Warm-Up or Daily Homework Quiz: TE pages 407 and 406, CRB page 23, or Transparencies

TEACHING OPTIONS

____ Lesson Opener (Activity): CRB page 24 or Transparencies
____ Examples: 2–9: SE pages 407–410
____ Extra Examples: TE pages 408–410 or Transparencies; Internet
____ Closure Question: TE page 410
____ Guided Practice Exercises: SE page 411

APPLY/HOMEWORK

Homework Assignment

____ Block Schedule: 22–84 even, 98, 101–111 odd; Quiz 1: 1–22

Reteaching the Lesson

____ Practice Masters: CRB pages 25–27 (Level A, Level B, Level C)
____ Reteaching with Practice: CRB pages 28–29 or Practice Workbook with Examples
____ Personal Student Tutor

Extending the Lesson

____ Applications (Real-Life): CRB page 31
____ Challenge: SE page 413; CRB page 32 or Internet

ASSESSMENT OPTIONS

____ Checkpoint Exercises: TE pages 408–410 or Transparencies
____ Daily Homework Quiz (7.2): TE page 414, CRB page 36, or Transparencies
____ Standardized Test Practice: SE page 413; TE page 414; STP Workbook; Transparencies
____ Quiz (7.1–7.2): SE page 414; CRB page 33

Notes ______________________________

NAME ______________________ DATE ________

Available as a transparency

Warm-Up Exercises

For use before Lesson 7.2, pages 407–414

Evaluate the expression.

1. $4^2 \cdot 4^3$

2. $(2^2)^3$

3. $\dfrac{3^4}{3^2}$

4. 2^{-3}

5. $(2 \cdot 3)^4$

Daily Homework Quiz

For use after Lesson 7.1, pages 401–406

Evaluate the expression.

1. $\left(\sqrt[4]{24}\right)^2$

2. $9^{5/4}$

3. $16^{-1/4}$

Evaluate the expression using a calculator. Round the result to two decimal places.

4. $\sqrt[3]{225}$

5. $12^{1/4}$

Solve the equation. Round the result to two decimal places.

6. $2x^4 = 35$

7. $5x^3 + 10 = 961$

NAME ______________________ DATE ________

Available as a transparency

Activity Lesson Opener

For use with pages 407–414

SET UP: Work in groups of 2 to 4.

YOU WILL NEED: • Number cube • Pencil and paper

Use the rules below to play the Power Game!

- Take turns; the winner is the first player to get a total of at least 60 points.
- Roll the number cube and choose an expression from the appropriate column. (For example, if you roll a 2, use Column 2.)
- Simplify your expression. The *exponent* tells you how many points you receive. So, you would get 7 points if your simplified expression is c^7, $3p^7$, or $\frac{2}{x^7}$.
- Each expression can be used only once. If (and *only* if) you roll a number for which the expressions are all used up, take the number you rolled as the number of points.
- Watch your opponents to make sure they don't make math errors! The penalty for a math error is 10 points.

1	2	3	4	5	6
$(n^4)^2$	$4u^{-7}$	$(2x)^5$	$(10m)^2$	$(z^2)^{-3}$	$(t^2)^3$
$(4d)^3$	$a^3 \cdot a$	$2c^{-8}$	$x^3 \cdot x^5$	y^{-3}	$e^4 \cdot e^7$
$j^2 \cdot j^3$	$(3h)^5$	$(u^2)^3$	g^{-4}	$(v^5)^2$	$(7q)^3$
$\frac{k^3}{k^7}$	$\left(\frac{5}{z^2}\right)^3$	$\frac{t^3}{t^5}$	$\left(\frac{u^3}{7}\right)^4$	$\frac{p^{10}}{p^8}$	$\frac{x}{x^5}$

LESSON 7.2

NAME _______________ DATE _______

Practice A

For use with pages 407–414

Simplify the expression using the properties of rational exponents.

1. $4^{1/3} \cdot 4^{4/3}$
2. $(6^{3/4})^{1/3}$
3. $(5 \cdot 3)^{2/3}$
4. $13^{-5/4}$
5. $\frac{10^{5/6}}{10^{4/6}}$
6. $\left(\frac{2}{3}\right)^{1/8}$

Simplify the expression using the properties of radicals.

7. $\sqrt{7} \cdot \sqrt{3}$
8. $\sqrt{\sqrt[3]{5}}$
9. $\frac{\sqrt{20}}{\sqrt{5}}$
10. $\sqrt{\frac{1}{25}}$
11. $\sqrt[5]{\sqrt{3}}$
12. $\frac{\sqrt{6} \cdot \sqrt{30}}{\sqrt{10}}$

Simplify the expression. Assume all variables are positive.

13. $x^{1/4} \cdot x^{2/4}$
14. $(x^{2/3})^4$
15. $(x^{3/5})^{1/6}$
16. $(3x)^{1/2}$
17. $(27x)^{1/3}$
18. $x^{-7/2}$
19. $\frac{x^{5/3}}{x^{2/3}}$
20. $\frac{x^{1/2}}{x^{5/2}}$
21. $\left(\frac{100}{x}\right)^{1/2}$

Perform the indicated operation. Assume all variables are positive.

22. $2\sqrt{3} + 4\sqrt{3}$
23. $5\sqrt{7} - 3\sqrt{7}$
24. $6\sqrt[5]{22} + 9\sqrt[5]{22}$
25. $2\sqrt{x} - 7\sqrt{x}$
26. $-5\sqrt[3]{x} + 2\sqrt[3]{x}$
27. $8\sqrt[4]{x} - 6\sqrt[4]{x}$

Write the expression in simplest form. Assume all variables are positive.

28. $\sqrt{49x^3}$
29. $\sqrt{\frac{x^2}{y^3}}$
30. $\sqrt[4]{x^3y^5z^8}$
31. $\sqrt[3]{x^3yz} + \sqrt[3]{8x^3yz}$
32. $\sqrt{x^5} + 5x\sqrt{x^3}$
33. $\frac{\sqrt{x^2y^4z}}{\sqrt{x^5}}$

34. ***Geometry*** The area of an equilateral triangle is given by $A = \frac{\sqrt{3}}{4}s^2$.

    Find the length of the side s of an equilateral triangle with an area of $\sqrt{12}$ square inches.

Lesson 7.2

LESSON 7.2

NAME ______________________________ DATE ____________

Practice B

For use with pages 407–414

Simplify the expression using the properties of radicals and rational exponents.

1. $5^{2/3} \cdot 5^{4/3}$
2. $\frac{3^{1/2}}{3}$
3. $(7^{2/3})^{5/2}$
4. $3^{1/4} \cdot 4^{1/4}$
5. $\sqrt[3]{2} \cdot \sqrt[3]{4}$
6. $\frac{\sqrt[4]{240}}{\sqrt[4]{15}}$
7. $\frac{\sqrt[3]{3}}{3}$
8. $\left(\frac{64}{125}\right)^{1/3}$
9. $(10^{1/2})^{2/3}$

Simplify the expression. Assume all variables are positive.

10. $\sqrt{9x^2}$
11. $\sqrt[3]{2x^3}$
12. $x^{2/3} \cdot x^{1/3}$
13. $\left(\frac{x}{4}\right)^{1/2}$
14. $(16x)^{1/4}$
15. $\sqrt[5]{27x} \cdot \sqrt[5]{9x^4}$
16. $\frac{\sqrt{12x^2}}{\sqrt{3}}$
17. $\frac{1}{(x^2)^{-1/3}}$
18. $\sqrt[4]{256xy^4}$
19. $x^3 \cdot x^{\sqrt{5}}$
20. $\left(x^{\sqrt{3}}\right)^{\sqrt{3}}$
21. $(4x)^{\sqrt{2}}$
22. $\frac{x^{\sqrt{3}}}{x^{5\sqrt{3}}}$
23. $x^{-\sqrt{2}}$
24. $3x^{\sqrt{6}} - 2x^{\sqrt{6}}$

Perform the indicated operation.

25. $2\sqrt[3]{3} + \sqrt[3]{3}$
26. $-3\sqrt[4]{15} + 2\sqrt[4]{15}$
27. $3(2^{1/3}) + 5(2^{1/3})$
28. $4\sqrt{2} - \sqrt{8}$
29. $\sqrt[3]{40} + \sqrt[3]{5}$
30. $\sqrt[5]{96} - 4\sqrt[5]{3}$
31. ***Milky Way*** The Milky Way is 10^5 light years in diameter and 10^4 light years in thickness. One light year is equivalent to 5.88×10^{12} miles. What is the diameter and thickness of the Milky Way in miles?

Archery Target **In Exercises 32–34, use the following information.**

The figure at the right shows a National Field Archer's Association official hunter's target. The area of the entire hunter's target is approximately 490.9 square inches. The area of the center white circle is approximately 19.6 square inches.

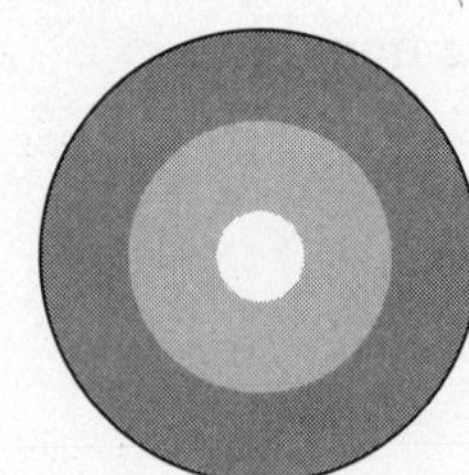

32. Find the radius of the target.
33. Find the radius of the center white circle.
34. Find the ratio of the radius of the white circle to the radius of the target.

Lesson 7.2

LESSON **7.2**

NAME ____ DATE ____________

Practice C

For use with pages 407–414

Simplify the expressions using the properties of radicals and rational exponents.

1. $(3^{1/2} \cdot 5^{2/3})^{3/2}$
2. $(2^{1/3} \cdot 2^{3/4})^{1/2}$
3. $((5^{2/3})^{1/5})^2$
4. $\left(\frac{6^{1/2}}{6^{1/3}}\right)^{3/5}$
5. $\left(\frac{3^{1/2}}{12^{1/2}}\right)^3$
6. $\sqrt[4]{\sqrt[3]{\sqrt{2}}}$
7. $\sqrt{\frac{\sqrt{108}}{\sqrt{27}}}$
8. $\sqrt[7]{(2^2)^3 \cdot (2^2)^4}$
9. $\frac{\sqrt{\frac{3}{18}} \cdot \sqrt{3}}{\sqrt{5}}$

Simplify the expression. Assume all the variables are positive.

10. $\frac{x^{5/4}\, y^{2/3}}{xy}$
11. $\left(\frac{3x^{1/4}\, y^{2/3} z}{2xy^{1/2}}\right)^2$
12. $\left(\frac{x^{4/3} y^5}{16z^{1/2}}\right)^{-1/4}$
13. $\left[\frac{(3xy)^{1/2}}{(27x^2y)^{1/2}}\right]^{-1}$
14. $\sqrt[5]{(2x^2)^3(2x^2)^7}$
15. $\sqrt[3]{\sqrt[4]{x}} \cdot \sqrt{\sqrt[3]{x}}$

Perform the indicated operations. Assume all variables are positive.

16. $\sqrt{(4\sqrt{6} - 3\sqrt{6})}$
17. $\sqrt[3]{8x^6y^2z} + x\sqrt[3]{27x^3y^2z}$
18. $\sqrt{-\sqrt[5]{x} + \sqrt[5]{32x}}$
19. $\sqrt[3]{\frac{5}{y}} + \sqrt{\frac{9}{y^2}}$
20. $\sqrt[5]{y} + \sqrt[10]{y^2} - 3\sqrt[15]{y^3}$
21. $\sqrt{xy}\sqrt{3x^2y}\sqrt{3xy} - \sqrt{6x^3y}\sqrt{x}\sqrt{54y^2}$

Halley's Comet **In Exercises 22 and 23, use the following information.**

Halley's Comet travels in an elliptical orbit around the sun, making one complete orbit every 76 years. When the comet was closest to the sun (8.9×10^{10} meters), it developed its tail. In the diagram at the right, a is the length of the semi-major axis, A is the comet's closest distance to the sun, and B is the comet's farthest distance from the sun.

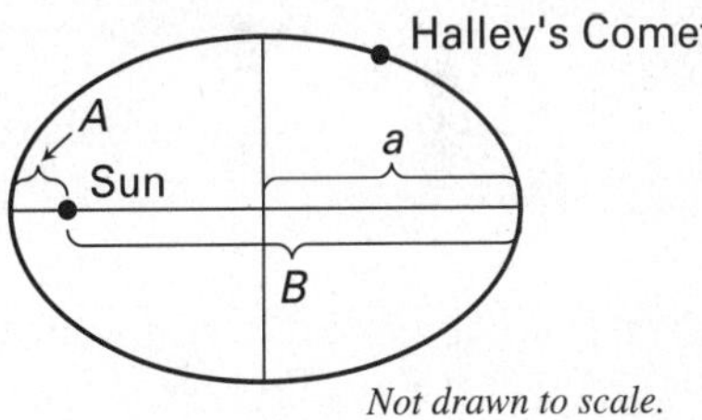

Not drawn to scale.

22. The length of the semi-major axis a can be found by the equation $a = \left(\frac{GMT^2}{4\pi^2}\right)^{1/3}$ where

    G = gravitational constant = 6.67×10^{-11} N · m²/kg²

    M = mass of sun = 1.99×10^{30} kg

    T = period = 2.4×10^9 seconds (76 years).

    Find the length of the semi-major axis.

23. The comet's farthest distance from the sun can be calculated by $B = 2a - A$. What's the comet's farthest distance from the sun?

NAME ______________________ DATE __________

Reteaching with Practice

For use with pages 407–414

GOAL **Use properties of rational exponents to evaluate and simplify expressions**

VOCABULARY

For a radical to be in **simplest form,** you must apply the properties of radicals, remove any perfect *n*th powers (other than 1) and rationalize any denominators. Two radical expressions are **like radicals** if they have the same index and the same radicand.

EXAMPLE 1 *Using Properties of Rational Exponents*

Use the properties of rational exponents to simplify the expression.

a. $x^{2/3} \cdot x^{1/9} = x^{(2/3+1/9)} = x^{(6/9+1/9)} = x^{7/9}$

b. $(2^3x^6)^{1/3} = (2^3)^{1/3} \cdot (x^6)^{1/3} = 2^{(3 \cdot 1/3)} \cdot x^{(6 \cdot 1/3)} = 2^1 \cdot x^2 = 2x^2$

c. $\dfrac{x^{3/4}}{x} = x^{(3/4-1)} = x^{(3/4-4/4)} = x^{-1/4} = \dfrac{1}{x^{1/4}}$

Exercises for Example 1

Use the properties of rational exponents to simplify the expression.

1. $x \cdot x^{1/2}$

2. $y^{-2/3}$

3. $(4^{2/3})^6$

4. $\dfrac{y^{2/3}}{y^{1/3}}$

5. $\dfrac{1}{64^{-1/3}}$

6. $z^{2/3} \cdot z^{1/2}$

EXAMPLE 2 *Using Properties of Radicals*

Use the properties of radicals to simplify the expression.

a. $\sqrt{8} \cdot \sqrt{2} = \sqrt{8 \cdot 2} = \sqrt{16} = 4$ Use the product property.

b. $\dfrac{\sqrt[3]{320}}{\sqrt[3]{5}} = \sqrt[3]{\dfrac{320}{5}} = \sqrt[3]{64} = 4$ Use the quotient property.

Exercises for Example 2

Use the properties of radicals to simplify the expression.

7. $\sqrt[3]{16} \cdot \sqrt[3]{4}$

8. $\sqrt{6} \cdot \sqrt{6}$

9. $\dfrac{\sqrt[4]{32}}{\sqrt[4]{2}}$

10. $\dfrac{\sqrt[3]{250}}{\sqrt[3]{2}}$

EXAMPLE 3 *Writing Radicals and Variable Expressions in Simplest Form*

Write the expression in simplest form. Assume all variables are positive.

a. $\sqrt[4]{96} = \sqrt[4]{16 \cdot 6}$ Factor out perfect fourth power.

$= \sqrt[4]{16} \cdot \sqrt[4]{6}$ Use the product property.

$= 2\sqrt[4]{6}$ Simplify $\sqrt[4]{16} = \sqrt[4]{2^4} = 2$.

NAME ______________________ DATE ___________

Reteaching with Practice

For use with pages 407–414

b. $\sqrt[3]{\dfrac{12}{25}} = \sqrt[3]{\dfrac{12 \cdot 5}{25 \cdot 5}}$ Make the denominator a perfect cube.

$= \sqrt[3]{\dfrac{60}{125}}$ Simplify.

$= \dfrac{\sqrt[3]{60}}{\sqrt[3]{125}}$ Quotient property

$= \dfrac{\sqrt[3]{60}}{5}$ Simplify $\sqrt[3]{125} = \sqrt[3]{5^3} = 5$.

c. $\sqrt{28y^5} = \sqrt{2^2 \cdot 7y^4y}$ Factor out perfect square.

$= \sqrt{2^2y^4} \cdot \sqrt{7y}$ Product property

$= 2y^2\sqrt{7y}$ Simplify $\sqrt{2^2(y^2)^2} = 2y^2$.

d. $\sqrt[5]{\dfrac{x^5}{y^3}} = \sqrt[5]{\dfrac{x^5y^2}{y^3y^2}}$ Make the denominator a perfect fifth power.

$= \dfrac{\sqrt[5]{x^5y^2}}{\sqrt[5]{y^5}}$ Simplify and use quotient property.

$= \dfrac{x\sqrt[5]{y^2}}{y}$ Simplify.

Exercises for Example 3

Write the expression in simplest form. Assume all variables are positive.

11. $\sqrt[3]{32}$ **12.** $\sqrt[4]{\dfrac{2}{9}}$ **13.** $\sqrt[4]{256x^8y}$ **14.** $\sqrt{\dfrac{4x^2y}{9z^2}}$

EXAMPLE 4 Adding and Subtracting Roots, Radicals, and Variable Expressions

Perform the indicated operation. Assume all variables are positive.

a. $4(3)^{1/3} - 2(3)^{1/3} = (4 - 2)(3)^{1/3} = 2(3)^{1/3}$

b. $\sqrt{27} + \sqrt{12} = \sqrt{9 \cdot 3} + \sqrt{4 \cdot 3}$ Factor out perfect squares.

$= \sqrt{9} \cdot \sqrt{3} + \sqrt{4} \cdot \sqrt{3}$ Product property

$= 3\sqrt{3} + 2\sqrt{3}$ Simplify.

$= (3 + 2)\sqrt{3}$ Distributive property

$= 5\sqrt{3}$ Simplify.

c. $\sqrt[3]{y} + 4\sqrt[3]{y} = (1 + 4)\sqrt[3]{y} = 5\sqrt[3]{y}$

Exercises for Example 4

Perform the indicated operation. Assume all variables are positive.

15. $2\sqrt[5]{3} - \sqrt[5]{3}$ **16.** $7(2^{1/8}) + 4(2^{1/8})$ **17.** $4\sqrt{x} + 2\sqrt{x}$

LESSON 7.2

NAME ______________________________ DATE ________

Quick Catch-Up for Absent Students

For use with pages 407–414

The items checked below were covered in class on (date missed) ___________

Lesson 7.2: Properties of Rational Exponents

____ **Goal 1:** Use properties of rational exponents and radicals to evaluate and simplify expressions. (pp. 407–409)

Material Covered:

____ Student Help: Look Back

____ Example 1: Using Properties of Rational Exponents

____ Example 2: Using Properties of Radicals

____ Example 3: Writing Radicals in Simplest Form

____ Example 4: Adding and Subtracting Roots and Radicals

____ Example 5: Simplifying Expressions Involving Variables

____ Example 6: Writing Variable Expressions in Simplest Form

____ Example 7: Adding and Subtracting Expressions Involving Variables

Vocabulary:

simplest form, p. 408 like radicals, p. 408

____ **Goal 2:** Use properties of rational exponents to solve real-life problems. (p. 410)

Material Covered:

____ Example 8: Evaluating a Model Using Properties of Rational Exponents

____ Example 9: Using Properties of Rational Exponents with Variables

____ Other (specify) ______________________________

____

Homework and Additional Learning Support

____ Textbook (specify) pp. 411–414

____

____ Internet: Extra Examples at www.mcdougallittell.com

____ *Reteaching with Practice* worksheet (specify exercises) ______________

____ *Personal Student Tutor* for Lesson 7.2

Lesson 7.2

LESSON 7.2

NAME ______________________ DATE __________

Real-Life Application: When Will I Ever Use This?

For use with pages 407–414

Fluorescent Light Bulbs

Fluorescent light bulbs are found in offices, schools, and factories. Fluorescent lights are long white tubes and are usually used in pairs, 2 or 4 at a time. Fluorescent light bulbs are different from incandescent light bulbs which are usually used in houses. A fluorescent lamp uses about 80% less electricity and produces 80% less heat as an incandescent lamp.

The fluorescent light bulb was introduced during the World's Fair in 1938–1939. Since 1952, the sales of fluorescent bulbs has been higher than the sales of incandescent bulbs. In 1938, one fluorescent light bulb was expected to last 1500 hours. Today, one florescent light bulb can last approximately 100,000 hours.

1. Find the number of days a florescent light bulb is expected to last when a bulb is expected to last $2^{16.61}$ hours and the bulb is used $2^{3.59}$ hours per day.
2. A florescent light bulb is expected to last $2^{16.61}$ hours and you have 2^2 bulbs. Find the total number of hours the bulbs are expected to last.
3. Using the expression $\left(\frac{28.85}{20.8}\right)^3$, find the number of fluorescent light bulbs used in 1974 for every one bulb used in 1989.
4. Using the expression $\left(\frac{10}{7.5}\right)^5$, find the number of fluorescent light bulbs used in 1989 for every one bulb used in 2000.

Lesson 7.2

LESSON 7.2

NAME ______________________ DATE __________

Challenge: Skills and Applications

For use with pages 407–414

In Exercises 1–6, simplify the expression.

1. $\sqrt[12]{3} \cdot \sqrt[12]{27}$

2. $\sqrt{\sqrt{8} \cdot \sqrt[4]{2}}$

3. $\sqrt[5]{\sqrt[6]{10} \cdot \sqrt{1000}}$

4. $\dfrac{\sqrt[3]{4}}{\sqrt[6]{2}}$

5. $\dfrac{\sqrt[3]{25}}{\sqrt{5}}$

6. $\sqrt[6]{9} \cdot \sqrt[3]{81}$

In Exercises 7–9, solve the equation.

7. $2^x \cdot 2^{x+1} = \dfrac{1}{8}$

8. $\dfrac{3x^2}{9^x} = 27$

9. $(2^x)^{x+1} = 64$

10. A good approximation of $(1 + x)^n$ is given by $(1 + x)^n \approx 1 + nx$, when x is a number close to 0.

a. Give an approximation of $(1 + x)^{m+n}$ for two real numbers m and n, using the formula above.

b. Find an approximation of $(1 + x)^{m+n}$, using the fact that $(1 + x)^{m+n} = (1 + x)^m \cdot (1 + x)^n$.

c. The two approximations you found in parts (a) and (b) differ by one term. Explain why this term is negligible, assuming, for example, that $x \approx 0.01$.

11. The French mathematician Francois Viéte (1540–1603) discovered a formula that expresses π as an "infinite product" of square roots:

$$\frac{2}{\pi} = \frac{\sqrt{2}}{2} \cdot \frac{\sqrt{2 + \sqrt{2}}}{2} \cdot \frac{\sqrt{2 + \sqrt{2 + \sqrt{2}}}}{2} \cdots$$

a. Verify that each factor f_n of this product $(n \geq 2)$ can be found from the formula

$$f_n = \sqrt{\frac{1 + f_{n-1}}{2}},$$

and use this formula to find the next factor not shown. $\left(\textit{Hint:} \text{ Write } f_{n-1} \text{ as } \frac{w}{2}.\right)$

b. Use the first 4 factors of Viéte's product to approximate π.

Lesson 7.2

LESSON 7.2

NAME ______________________ DATE ____________

Quiz 1

For use after Lessons 7.1–7.2

Evaluate the expression without using a calculator. *(Lesson 7.1)*

1. $16^{\frac{3}{4}}$
2. $25^{-\frac{3}{2}}$
3. $-32^{\frac{4}{5}}$
4. $(-8)^{\frac{2}{3}}$

Solve the equation. Round your answer to two decimal places. *(Lesson 7.1)*

5. $2x^2 = 12$
6. $(x + 3)^3 = 14$

Write the expression in simplest form. *(Lesson 7.2)*

7. $\dfrac{1}{3^{-\frac{1}{5}}}$
8. $\sqrt[3]{\dfrac{8}{3}}$
9. $\sqrt[3]{3} \cdot \sqrt{9}$
10. $2^{\frac{2}{3}} + 3\left(2^{\frac{2}{3}}\right)$

Write the expression in simplest form. Assume all variables are positive. *(Lesson 7.2)*

11. $\sqrt[3]{x^2} \cdot \sqrt{x}$
12. $\left(x^{\frac{2}{3}}\right)^{\frac{3}{4}}$
13. $\sqrt[3]{81x^5y^3}$
14. $\dfrac{x^{\frac{2}{3}}y}{x^{-2}y^{\frac{3}{4}}}$
15. The volume of a sphere can be found using the formula $V = \frac{4}{3}\pi r^3$ where r is the radius of the sphere. Find the length of the radius of a sphere that has a volume of 57.9 m³. *(Lesson 7.2)*

Answers

1. ____________
2. ____________
3. ____________
4. ____________
5. ____________
6. ____________
7. ____________
8. ____________
9. ____________
10. ____________
11. ____________
12. ____________
13. ____________
14. ____________
15. ____________

Lesson 7.2

LESSON 7.3

TEACHER'S NAME ______________________ CLASS ______ ROOM ______ DATE ______

Lesson Plan

1-day lesson (See *Pacing the Chapter,* TE pages 398C–398D) **For use with pages 415–420**

GOALS

1. **Perform operations with functions including power functions.**
2. **Use power functions and function operations to solve real-life problems.**

State/Local Objectives ____

____

✓ Check the items you wish to use for this lesson.

STARTING OPTIONS

____ Homework Check: TE page 411; Answer Transparencies
____ Warm-Up or Daily Homework Quiz: TE pages 415 and 414, CRB page 36, or Transparencies

TEACHING OPTIONS

____ Lesson Opener (Application): CRB page 37 or Transparencies
____ Graphing Calculator Activity with Keystrokes: CRB pages 38–39
____ Examples 1–5: SE pages 415–417
____ Extra Examples: TE pages 416–417 or Transparencies
____ Closure Question: TE page 417
____ Guided Practice Exercises: SE page 418

APPLY/HOMEWORK

Homework Assignment

____ Basic 12–36 even, 40–48 even, 52, 57–61, 69–83 odd
____ Average 12–50 even, 52, 53, 57–61, 69–83 odd
____ Advanced 12–50 even, 52–54, 56–67, 69–83 odd, 84

Reteaching the Lesson

____ Practice Masters: CRB pages 40–42 (Level A, Level B, Level C)
____ Reteaching with Practice: CRB pages 43–44 or Practice Workbook with Examples
____ Personal Student Tutor

Extending the Lesson

____ Applications (Interdisciplinary): CRB page 46
____ Challenge: SE page 420; CRB page 47 or Internet

ASSESSMENT OPTIONS

____ Checkpoint Exercises: TE pages 416–417 or Transparencies
____ Daily Homework Quiz (7.3): TE page 420, CRB page 50, or Transparencies
____ Standardized Test Practice: SE page 420; TE page 420; STP Workbook; Transparencies

Notes ____

____

____

LESSON

7.3

TEACHER'S NAME _______________ CLASS _______ ROOM _______ DATE _______

Lesson Plan for Block Scheduling

Half-day lesson (See *Pacing the Chapter,* TE pages 398C–398D) **For use with pages 415–420**

1. **Perform operations with functions including power functions.**
2. **Use power functions and function operations to solve real-life problems.**

State/Local Objectives _______________

____

____

CHAPTER PACING GUIDE	
Day	**Lesson**
1	Assess Ch. 6; 7.1(all)
2	7.2 (all)
3	**7.3 (all)**; 7.4(begin)
4	7.4 (end); 7.5(all)
5	7.6 (all)
6	7.7 (all)
7	Review/Assess Ch. 7

✓ Check the items you wish to use for this lesson.

STARTING OPTIONS

____ Homework Check: TE page 411; Answer Transparencies
____ Warm-Up or Daily Homework Quiz: TE pages 415 and 414, CRB page 36, or Transparencies

TEACHING OPTIONS

____ Lesson Opener (Application): CRB page 37 or Transparencies
____ Graphing Calculator Activity with Keystrokes: CRB pages 38–39
____ Examples 1–5: SE pages 415–417
____ Extra Examples: TE pages 416–417 or Transparencies
____ Closure Question: TE page 417
____ Guided Practice Exercises: SE page 418

APPLY/HOMEWORK

Homework Assignment (See also the assignment for Lesson 7.4.)

____ Block Schedule: 12–50 even, 52–54, 56–67, 69–83 odd, 84

Reteaching the Lesson

____ Practice Masters: CRB pages 40–42 (Level A, Level B, Level C)
____ Reteaching with Practice: CRB pages 43–44 or Practice Workbook with Examples
____ Personal Student Tutor

Extending the Lesson

____ Applications (Interdisciplinary): CRB page 46
____ Challenge: SE page 420; CRB page 47 or Internet

ASSESSMENT OPTIONS

____ Checkpoint Exercises: TE pages 416–417 or Transparencies
____ Daily Homework Quiz (7.3): TE page 420, CRB page 50, or Transparencies
____ Standardized Test Practice: SE page 420; TE page 420; STP Workbook; Transparencies

Notes _______________

____

____

LESSON

7.3

NAME ______________________ DATE __________

Available as a transparency

WARM-UP EXERCISES

For use before Lesson 7.3, pages 415–420

Simplify.

1. $4(x^2 + 1)$
2. $(x - 2) + (x^2 + 1)$
3. $(x - 2) - (x^2 + 1)$
4. $(x - 2)(x + 1)$
5. $(x - 2)^2$

DAILY HOMEWORK QUIZ

For use after Lesson 7.2, pages 407–414

Simplify the expression. Assume all variables are positive.

1. $6^{3/8} \cdot 6^{5/8}$
2. $\sqrt[6]{17} \cdot \sqrt[3]{17}$
3. $\sqrt[4]{5} \cdot \sqrt[4]{125}$
4. $y^{3/2} \cdot y^{1/3}$
5. $\sqrt{121z^5}$

Perform the indicated operation.

6. $2\sqrt[3]{4} + 3\sqrt[3]{4}$

LESSON 7.3

NAME ______________________ DATE __________

Available as a transparency

Application Lesson Opener

For use with pages 415–420

Suppose the function $p(x) = -0.02x^2 + 45x - 4000$ gives the profit that a store obtains by selling x bicycles in one month. You can substitute various values or expressions for x in order to find the profit. Here are some examples.

$$p(600) = -0.02(600)^2 + 45(600) - 4000 = 15{,}800$$

$$p(t) = -0.02t^2 + 45t - 4000$$

$$p(x + 10) = -0.02(x + 10)^2 + 45(x + 10) - 4000$$

$$= -0.02t^2 + 44.6x - 3552$$

The function $f(x) = 14x$ gives the total cost of buying x CDs at $14 each. Evaluate each of the following.

1. $f(3)$

2. $f(t)$

3. $f(3s)$

4. $f(2x + 5)$

The function $g(x) = \dfrac{300}{x}$ gives the number of hours required to drive 300 miles at a speed of x miles per hour. Evaluate each of the following.

5. $g(50)$

6. $g(r)$

7. $g(20p)$

8. $g(75 - x)$

The function $h(t) = -16t^2 + 48t$ gives the height in feet after t seconds of a projectile fired upward from ground level at a speed of 48 feet per second. Evaluate each of the following.

9. $h(2.5)$

10. $h(x)$

11. $h(3x)$

12. $h(t + 1)$

LESSON

7.3

NAME ______________________ DATE __________

Graphing Calculator Activity

For use with pages 415–420

GOAL **To explore the function operations of addition, subtraction, and multiplication**

Let f and g be any two functions. A new function h can be defined by performing any of the four basic operations (addition, subtraction, multiplication, and division) on f and g.

For example, let $f(x) = x + 5$ and $g(x) = 3x$.

Then $f(x) + g(x) = x + 5 + 3x = 4x + 5$

$f(x) - g(x) = (x + 5) - 3x = -2x + 5$

$f(x) \cdot g(x) = (x + 5)(3x) = 3x^2 + 15x$

Activity

Let $f(x) = 2x + 1$ and $g(x) = x - 2$

1. In order to verify that $f(x) + g(x) = 3x - 1$, graph both $y = (2x + 1) + (x - 2)$ and $y = 3x - 1$.

Notice that the graphs coincide.

2. In order to verify that $f(x) - g(x) = x - 3$, graph both $y = (2x + 1) - (x - 2)$ and $y = x - 3$.

Notice that the graphs coincide.

3. In order to verify that $f(x) \cdot g(x) = (2x + 1)(x - 2) = 2x^2 - 3x - 2$, graph both $y = (2x + 1)(x - 2)$ and $y = 2x^2 - 3x - 2$.

Notice that the graphs coincide.

Exercises

For Exercises 1–3, let $f(x) = 3x + 2$ and $g(x) = 5x - 7$.

1. Find $f(x) + g(x)$. Verify your answer by graphing both your answer and $y = (3x + 2) + (5x - 7)$.

2. Find $f(x) - g(x)$. Verify your answer by graphing both your answer and $y = (3x + 2) - (5x - 7)$.

3. Find $f(x) \cdot g(x)$. Verify your answer by graphing both your answer and $y = (3x + 2) \cdot (5x - 7)$.

LESSON 7.3 CONTINUED

NAME ______________________________ DATE __________

Graphing Calculator Activity

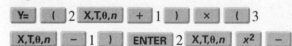

For use with pages 415–420

TI-82

Step 1

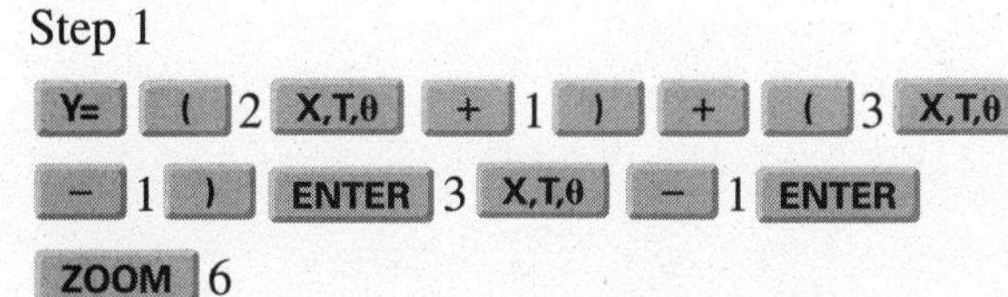

Step 2

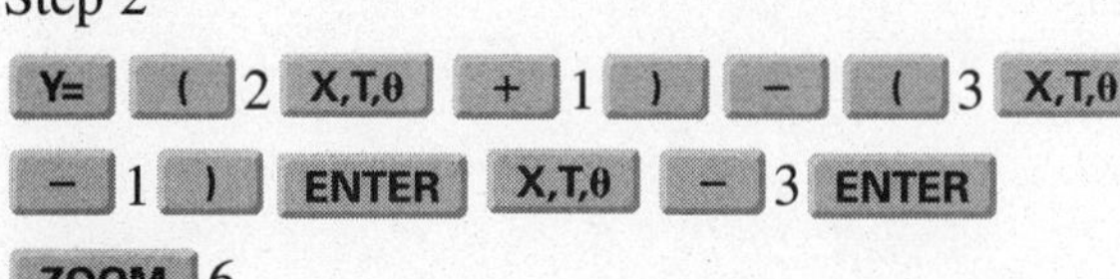

Step 3

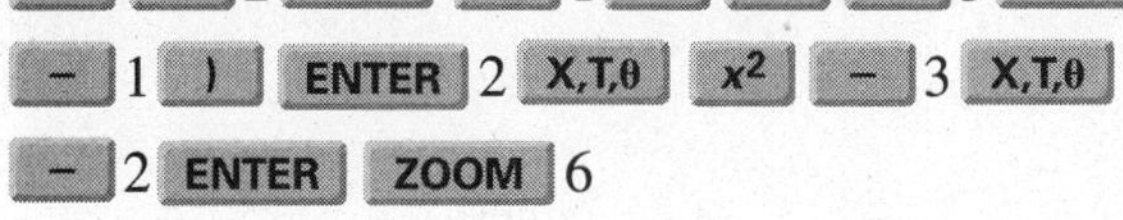

TI-83

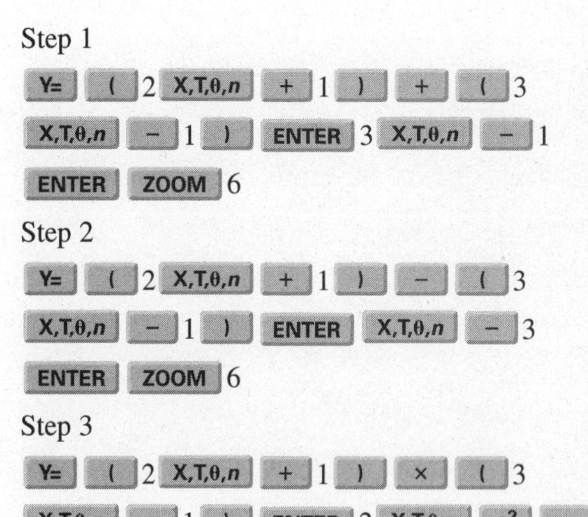

SHARP EL-9600C

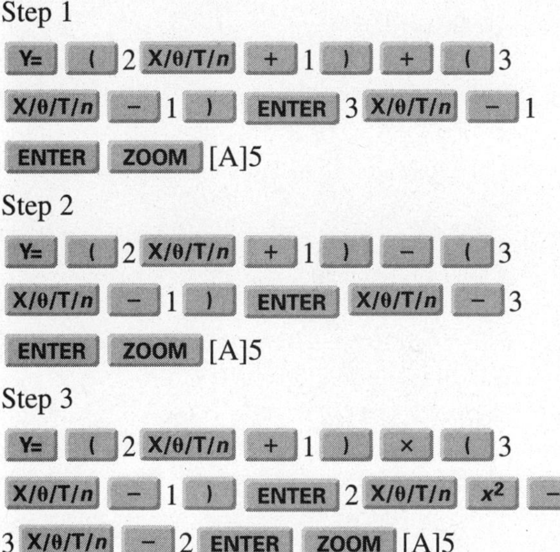

CASIO CFX-9850GA PLUS

From the main manu, choose GRAPH.

Step 1

(2 X,θ,T + 1) + (3 X,θ,T −
1) EXE 3 X,θ,T − 1 EXE SHIFT F3
F3 EXIT F6

Step 2

(2 X,θ,T + 1) − (3 X,θ,T −
1) EXE X,θ,T − 3 EXE SHIFT F3
F3 EXIT F6

Step 3

(2 X,θ,T + 1) × (3 X,θ,T −
1) EXE 2 X,θ,T x^2 − 3 X,θ,T −
2 EXE SHIFT F3 F3 EXIT F6

Lesson 7.3

LESSON 7.3

NAME ______________________ DATE ________

Practice A

For use with pages 415–420

Find $f(x)+g(x)$. Simplify your answer.

1. $f(x)=4x,\ g(x)=1-x$
2. $f(x)=2x+3,\ g(x)=x^2-1$
3. $f(x)=x^2+3,\ g(x)=x^2-2x-1$
4. $f(x)=x^{1/2},\ g(x)=6x^{1/2}$

Find $f(x)-g(x)$. Simplify your answer.

5. $f(x)=2x,\ g(x)=x+3$
6. $f(x)=x^2-x,\ g(x)=x^2-2$
7. $f(x)=x+1,\ g(x)=-x^2+2x+3$
8. $f(x)=3x^{3/2},\ g(x)=4x^{3/2}$

Find $f(x)\cdot g(x)$. Simplify your answer.

9. $f(x)=2x-1,\ g(x)=3$
10. $f(x)=x+1,\ g(x)=3x-2$
11. $f(x)=x^2+x-1,\ g(x)=2x$
12. $f(x)=2x^{2/3},\ g(x)=3x^{1/3}$

Find $\dfrac{f(x)}{g(x)}$. Simplify your answer.

13. $f(x)=3x,\ g(x)=x+2$
14. $f(x)=x^2+1,\ g(x)=x-2$
15. $f(x)=x-2,\ g(x)=x^2+x-4$
16. $f(x)=(2x)^{1/2},\ g(x)=2\sqrt{2}\,x^{1/3}$

Find $f(g(x))$. Simplify your answer.

17. $f(x)=2x,\ g(x)=x+5$
18. $f(x)=\sqrt{x},\ g(x)=4x+9$
19. $f(x)=x^2+2,\ g(x)=x-1$
20. $f(x)=x^{1/5},\ g(x)=x^{3/4}$

Let $f(x)=x^2$ and $g(x)=x-3$. Find the domain of the following functions.

21. $f(x)+g(x)$
22. $f(x)-g(x)$
23. $f(x)\cdot g(x)$
24. $\dfrac{f(x)}{g(x)}$
25. $f(g(x))$
26. $g(f(x))$
27. $g(x)-f(x)$
28. $\dfrac{g(x)}{f(x)}$
29. $f(f(x))$

30. ***Profit*** A company estimates that its cost and revenue can be modeled by the functions $C(x)=0.75x+20{,}000$ and $R(x)=1.50x$ where x is the number of units produced. The company's profit, P, is modeled by $P(x)=R(x)-C(x)$. Find the profit equation and determine the profit when 1,000,000 units are produced.

Lesson 7.3

LESSON

7.3

NAME ______________________ DATE __________

Practice B

For use with pages 415–420

Find $f(x) + g(x)$ and $f(x) - g(x)$. Simplify your answers.

1. $f(x) = 3x^3 - 2x^2 + 5x - 1,\ g(x) = x^2 + 7x - 1$
2. $f(x) = 4x^{2/3},\ g(x) = 3x^{2/3}$
3. $f(x) = 2x^3 - 3x + 4,\ g(x) = x^2 + 5x - 1$
4. $f(x) = \frac{1}{2}x^{3/4},\ g(x) = \frac{1}{8}x^{3/4}$

Find $f(x) \cdot g(x)$. Simplify your answer.

5. $f(x) = -x^2 + 2x + 2,\ g(x) = x + 1$
6. $f(x) = x^4 + 3x + 2,\ g(x) = x^2 + 3$
7. $f(x) = 2x^{1/4},\ g(x) = 2x^{1/3}$
8. $f(x) = 4x^{-1},\ g(x) = 2x^{1/2}$

Find $\frac{f(x)}{g(x)}$. Simplify your answer.

9. $f(x) = 3x^2 - x + 1,\ g(x) = x + 3$
10. $f(x) = 3x + 5,\ g(x) = 2x^2 - 1$
11. $f(x) = 6x^{7/3},\ g(x) = 3x^{2/3}$
12. $f(x) = (3x)^{1/4},\ g(x) = x^{5/4}$

Find $f(g(x))$ and $g(f(x))$. Simplify your answers.

13. $f(x) = 3x,\ g(x) = 2x + 1$
14. $f(x) = x^2 + 1,\ g(x) = x - 2$
15. $f(x) = -x^{1/2},\ g(x) = x + 4$
16. $f(x) = 3x^{4/5},\ g(x) = x^{1/2}$

Let $f(x) = 4x^{1/2}$ and $g(x) = x + 3$. Perform the given operation and state the domain.

17. $f(x) + g(x)$
18. $g(x) - f(x)$
19. $f(x) \cdot g(x)$
20. $\frac{g(x)}{f(x)}$
21. $f(g(x))$
22. $g(f(x))$

Furniture Sale **In Exercises 23–27, use the following information.** You have a coupon for \$100 off the price of a sofa. When you arrive at the store, you find that the sofas are on sale for 25% off. Let x represent the original price of the sofa.

23. Use function notation to describe your cost, $f(x)$, using only the coupon.
24. Use function notation to describe your cost, $g(x)$, with only the 25% discount.
25. Form the composition of the functions f and g that represents your cost, if you use the coupon first, then take the 25% discount.
26. Form the composition of the functions f and g that represents your cost if you use the discount first, then use the coupon.
27. Would you pay less for the sofa if you used the coupon first or took the 25% discount first?

LESSON 7.3

NAME ______________________ DATE __________

Practice C

For use with pages 415–420

Find $f(x) + g(x)$ and $f(x) - g(x)$. Simplify your answers.

1. $f(x) = x^3 + 3x^2 + 2x - 1,\ g(x) = x^5 - 2x^3 + 4x - 8$
2. $f(x) = 6x^{2/5} + 3x^{-1},\ g(x) = 4x^{2/5} - 5x^{-1}$
3. $f(x) = x^2 + 3x - 1,\ g(x) = 7x + 2$
4. $f(x) = 3x^{1/6} - 2x^3 - 1,\ g(x) = 2x^{1/6} - 5x^3$

Find $f(x) \cdot g(x)$. Simplify your answer.

5. $f(x) = x^3 + 2x^2 + x - 5,\ g(x) = x^2 + 2x - 6$
6. $f(x) = 5x^{1/4} - 3,\ g(x) = x^{3/8} - 1$

Find $\frac{f(x)}{g(x)}$. Simplify your answer.

7. $f(x) = 3x^{2/3} + 1,\ g(x) = x^{-1/3}$
8. $f(x) = 16x^{-1/3},\ g(x) = x^2$

Find $f(g(x))$ and $g(f(x))$. Simplify your answers.

9. $f(x) = (3 + x)^{1/2},\ g(x) = x^2 + 1$
10. $f(x) = x^{-2},\ g(x) = 3x - 1$
11. $f(x) = x^{3/4},\ g(x) = 2x$
12. $f(x) = 3x^{-1},\ g(x) = 2x^{1/2}$

Let $f(x) = x^{-1/2}$ and $g(x) = x^2 + 2x$. Perform the operation and state the domain.

13. $f(g(x))$
14. $g(f(x))$
15. $\frac{f(x)}{g(x)}$
16. $\frac{g(x)}{f(x)}$
17. $f(f(x))$
18. $g(g(x))$

Critical Thinking **State whether or not the following statements are always true. If they are false, give an example.**

19. $f(x) + g(x) = g(x) + f(x)$
20. $f(x) - g(x) = g(x) - f(x)$
21. $f(x) \cdot g(x) = g(x) \cdot f(x)$
22. $\frac{f(x)}{g(x)} = \frac{g(x)}{f(x)}$
23. $f(g(x)) = g(f(x))$
24. $f(f(x)) = [f(x)]^2$

Function Composition **Find functions f and g such that $h(x) = f(g(x))$.**

25. $h(x) = \sqrt{2x + 1}$
26. $h(x) = \frac{1}{3x + 2}$

27. ***Holiday Sale*** A department store is holding its annual end-of-year sale. Feature items are marked 40% off. In addition, a flyer was sent to the newspapers which included a coupon for \$5 off any purchase. Also, if you open a charge account with the store, you can receive an additional 10% discount. There are six different ways in which these price reductions can be composed. Find all six compositions. Which of the six compositions is the store most likely to use?

Lesson 7.3

NAME ______________________ DATE __________

Reteaching with Practice

For use with pages 415–420

GOAL **Perform operations with functions, including power functions**

VOCABULARY

A **power function** has the form $y = ax^b$, where a is a real number and b is a rational number.

The **composition** of the function f with the function g is given by $h(x) = f(g(x))$, where the domain of h is the set of all x-values such that x is in the domain of g, and $g(x)$ is in the domain of f.

EXAMPLE 1 Adding and Subtracting Functions

Let $f(x) = -2x$ and $g(x) = x + 3$. Perform the indicated operation and state the domain.

a. $f(x) + g(x) = -2x + (x + 3) = -x + 3$

b. $f(x) - g(x) = -2x - (x + 3) = -2x + (-x - 3) = -3x - 3$

The functions f and g each have the same domain–all real numbers. So, the domains of $f + g$ and $f - g$ also consist of all real numbers.

Exercises for Example 1

Let $f(x) = 2 - x$ and $g(x) = 3x$. Perform the indicated operation and state the domain.

1. $f(x) + g(x)$ **2.** $f(x) - g(x)$ **3.** $g(x) - f(x)$ **4.** $g(x) + g(x)$

EXAMPLE 2 Multiplying and Dividing Functions

Let $f(x) = 5x^3$ and $g(x) = x - 1$. Perform the indicated operation and state the domain.

a. $f(x) \cdot g(x) = 5x^3(x - 1) = (5x^3)(x) - (5x^3)(1) = 5x^4 - 5x^3$

The functions f and g each have the same domain-all real numbers. So, the domain of $f \cdot g$ also consists of all real numbers.

b. $\dfrac{f(x)}{g(x)} = \dfrac{5x^3}{x - 1}$

Since $x = 1$ will make the denominator zero, the domain is all real numbers except $x = 1$.

NAME ______________________ DATE __________

Reteaching with Practice

For use with pages 415–420

Exercises for Example 2

Perform the indicated operation and state the domain.

5. $f \cdot g; f(x) = x^{1/2}, g(x) = 3x^3$

6. $f \cdot g; f(x) = x + 3, g(x) = 2x^2$

7. $\frac{f}{g}; f(x) = 4x^{2/3}, g(x) = 2x$

8. $\frac{f}{g}; f(x) = -7x + 1, g(x) = x$

EXAMPLE 3 Finding the Composition of Functions

Let $f(x) = 3x - 2$ and $g(x) = x^2$. Find the following.

a. $f(g(x))$

b. $g(f(x))$

c. $f(f(x))$

d. the domain of each composition

SOLUTION

a. To find $f(g(x))$, substitute x^2 for x in the function f.

$f(g(x)) = f(x^2) = 3(x^2) - 2 = 3x^2 - 2$

b. To find $g(f(x))$, substitute $3x - 2$ for x in the function g.

$g(f(x)) = g(3x - 2) = (3x - 2)^2 = 9x^2 - 12x + 4$

c. To find $f(f(x))$, substitute $3x - 2$ for x in the function f.

$f(f(x) = f(3x - 2) = 3(3x - 2) - 2 = 9x - 6 - 2 = 9x - 8$

d. The functions f and g each have the same domain–all real numbers. So the domain of each composition also consists of all real numbers.

Exercises for Example 3

Let $f(x) = 2x^{-1}$ and $g(x) = x - 2$. Perform the indicated operation and state the domain.

9. $f(g(x))$

10. $g(f(x))$

11. $f(f(x))$

12. $g(g(x))$

LESSON 7.3

NAME ______________________ DATE ________

Quick Catch-Up for Absent Students

For use with pages 415–420

The items checked below were covered in class on (date missed) ___________

Lesson 7.3: Power Functions and Function Operations

____ **Goal 1:** Perform operations with functions including power functions. (pp. 415, 416)

Material Covered:

____ Example 1: Adding and Subtracting Functions

____ Student Help: Look Back

____ Example 2: Multiplying and Dividing Fractions

____ Student Help: Study Tip

____ Example 3: Finding the Composition of Functions

Vocabulary:

power function, p. 415 composition, p. 416

____ **Goal 2:** Use power functions and function operations to solve real-life problems. (p. 417)

Material Covered:

____ Example 4: Using Function Operations

____ Student Help: Skills Review

____ Example 5: Using Composition of Functions

____ Other (specify) ______________________

____

Homework and Additional Learning Support

____ Textbook (specify) pp. 418–420

____

____ *Reteaching with Practice* worksheet (specify exercises)______________

____ *Personal Student Tutor* for Lesson 7.3

NAME ______________________ DATE __________

Interdisciplinary Application

For use with pages 415–420

Waves

PHYSICS In Physics, a wave is a transfer of energy by the regular vibration of a physical medium or space. A wave felt during an earthquake travels through the earth, so the earth acts as a medium. For a wave made by using a rope, the rope is the medium. The medium for waves seen at a lake or ocean is the surface of the water.

Waves seen in water are classified as transverse waves. These types of waves move along as the medium moves up and down. These waves are called transverse because the motion of the waves is perpendicular to the motion of the medium. For example, if you stand in a body of water and you see a wave coming toward you, the wave is moving forward toward you, but you see the medium (the water) moving up and down. You can feel the medium moving up and down as the wave passes by you.

In Exercises 1–6, use the following information.

Your physics class decides to perform an experiment on circular waves. First, you make sure the water in the school swimming pool is level; that is, there are no waves. Next, a volunteer tosses a small marble near the center of the pool. Your teacher tells you the radius (in feet) of the outer ripple is given by $r(t) = 0.6t$, where t is the time in seconds after the marble hits the water. The area of the circle is given by the function $A(r) = \pi r^2$, where A is the area in square feet and r is the radius.

1. Find an equation for the composite function $A(r(t))$.
2. What is the input and output of the composite function?
3. Find the area of the circle 2 seconds after the marble hits the water.
4. Assume the marble hit exactly in the middle of the pool. The outer ripple takes 16.66 seconds to reach the long side of the pool m. Find the area of the circle.
5. Find the length of the short side of the pool ℓ.
6. The area of a circle formed by the outer ripple is 102 square feet. How long ago did the marble hit the water?

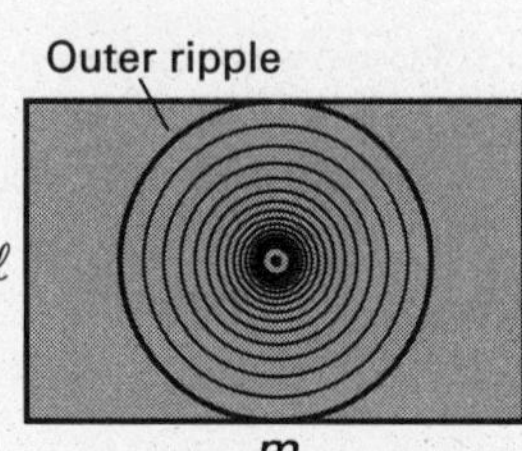

NAME ______________________ DATE __________

Challenge: Skills and Applications

For use with pages 415–420

For Exercises 1–4, use the following functions:

(i) $f(x) = x$ (ii) $f(x) = -x$ (iii) $f(x) = 1$

(iv) $f(x) = \frac{1}{x}$ (v) $f(x) = ix$ (vi) $f(x) = |x|$

Name *all* the functions (if any) in the list that satisfy each property.

1. $f(f(x)) = x$ **2.** $f(f(x)) = f(x)$

3. $f(f(x)) = -x$ **4.** $f(f(x)) = -f(x)$

For Exercises 5–7, suppose $f(x) = ax^m$ and $g(x) = bx^n$. Express each of the following in terms of *a, b, m, n,* and *x*.

5. $f(x) \cdot g(x)$ **6.** $f(g(x))$ **7.** $g(f(x))$

For Exercises 8–10, the notation $f^n(x)$ means the product of $f(x)$ with itself *n* times: $f(x) \cdot f(x) \cdot \cdots \cdot f(x)$. For each function express $f^n(x)$ explicitly in terms of *x* and *n*.

8. $f(x) = \sqrt{x}$ **9.** $f(x) = 3x^2$ **10.** $f(x) = \frac{2}{x}$

11. Suppose $f(x)$ and $g(x)$ are two functions with the same domain, and let $h(x) = f(x) + g(x)$. Suppose also that each of the 3 functions f, g and h, has a maximum value in this domain (i.e. a value that is greater than or equal to all the other values of the function). Let $M =$ the maximum value of $f(x)$, $N =$ the maximum value of $g(x)$, and $P =$ the maximum value of $h(x)$. Must it be true that $M + N = P$? Prove this, or state what relationship does exist between the numbers $M + N$ and P.

LESSON 7.4

TEACHER'S NAME ______ CLASS ______ ROOM ______ DATE ______

Lesson Plan

2-day lesson (See *Pacing the Chapter*, TE pages 398C–398D) **For use with pages 421–430**

1. Find inverses of linear functions.
2. Find inverses of nonlinear functions.

State/Local Objectives ______

✓ **Check the items you wish to use for this lesson.**

STARTING OPTIONS

____ Homework Check: TE page 418; Answer Transparencies
____ Warm-Up or Daily Homework Quiz: TE pages 422 and 420, CRB page 50, or Transparencies

TEACHING OPTIONS

____ Concept Activity: SE page 421; CRB page 51 (Activity Support Master)
____ Lesson Opener (Activity): CRB page 52 or Transparencies
____ Graphing Calculator Activity with Keystrokes: CRB page 53
____ Examples: Day 1: 1–4, SE pages 422–424; Day 2: 5–6, SE page 425
____ Extra Examples: Day 1: TE pages 423–424 or Transp.; Day 2: TE page 425 or Transp.
____ Technology Activity: SE page 430
____ Closure Question: TE page 425
____ Guided Practice: SE page 426 Day 1: Exs. 1–2, 4–10; Day 2: Exs. 3, 11–13

APPLY/HOMEWORK

Homework Assignment

____ Basic Day 1: 14–15, 16–30 even, 33–35, 36–38, 42; Day 2: 39–41, 44–56 even, 57, 59, 63–64, 69–75 odd; Quiz 2: 1–17
____ Average Day 1: 14–56 even, 57, 58, 62; Day 2: 33–55 odd, 58–62 even, 63–64, 69–75 odd; Quiz 2: 1–17
____ Advanced Day 1: 14–60 even; Day 2: 15–61 odd, 62–68, 69–83 odd; Quiz 2: 1–17

Reteaching the Lesson

____ Practice Masters: CRB pages 54–56 (Level A, Level B, Level C)
____ Reteaching with Practice: CRB pages 57–58 or Practice Workbook with Examples
____ Personal Student Tutor

Extending the Lesson

____ Applications (Real-Life): CRB page 60
____ Challenge: SE page 428; CRB page 61 or Internet

ASSESSMENT OPTIONS

____ Checkpoint Exercises: Day 1: TE pages 423–424 or Transp.; Day 2: TE page 425 or Transp.
____ Daily Homework Quiz (7.4): TE page 429, CRB page 65, or Transparencies
____ Standardized Test Practice: SE page 428; TE page 429; STP Workbook; Transparencies
____ Quiz (7.3–7.4): SE page 429; CRB page 62

Notes ______

LESSON 7.4

TEACHER'S NAME ______________ CLASS ______ ROOM ______ DATE ______

Lesson Plan for Block Scheduling

1-day lesson (See *Pacing the Chapter,* TE pages 398C–398D)

For use with pages 421–430

GOALS
1. Find inverses of linear functions.
2. Find inverses of nonlinear functions.

State/Local Objectives ______________

CHAPTER PACING GUIDE	
Day	**Lesson**
1	Assess Ch. 6; 7.1(all)
2	7.2 (all)
3	7.3 (all); **7.4(begin)**
4	**7.4 (end)**; 7.5(all)
5	7.6 (all)
6	7.7 (all)
7	Review/Assess Ch. 7

✓ **Check the items you wish to use for this lesson.**

STARTING OPTIONS

____ Homework Check: TE page 418; Answer Transparencies
____ Warm-Up or Daily Homework Quiz: TE pages 422 and 420, CRB page 50, or Transparencies

TEACHING OPTIONS

____ Concept Activity: SE page 421; CRB page 51 (Activity Support Master)
____ Lesson Opener (Activity): CRB page 52 or Transparencies
____ Graphing Calculator Activity with Keystrokes: CRB page 53
____ Examples: Day 3: 1–4, SE pages 422–424; Day 4: 5–6, SE page 425
____ Extra Examples: Day 3: TE pages 423–424 or Transp.; Day 4: TE page 425 or Transp.
____ Technology Activity: SE page 430
____ Closure Question: TE page 425
____ Guided Practice: SE page 426 Day 3: Exs. 1–2, 4–10; Day 4: Exs. 3, 11–13

APPLY/HOMEWORK

Homework Assignment (See also the assignments for Lessons 7.3 and 7.5.)

____ Block Schedule: Day 3: 14–56 even, 57, 58, 62; Day 4: 33–55 odd, 58–62 even, 63–64, 69–75 odd; Quiz 2: 1–17

Reteaching the Lesson

____ Practice Masters: CRB pages 54–56 (Level A, Level B, Level C)
____ Reteaching with Practice: CRB pages 57–58 or Practice Workbook with Examples
____ Personal Student Tutor

Extending the Lesson

____ Applications (Real-Life): CRB page 60
____ Challenge: SE page 428; CRB page 61 or Internet

ASSESSMENT OPTIONS

____ Checkpoint Exercises: Day 3: TE pages 423–424 or Transp.; Day 4: TE page 425 or Transp.
____ Daily Homework Quiz (7.4): TE page 429, CRB page 65, or Transparencies
____ Standardized Test Practice: SE page 428; TE page 429; STP Workbook; Transparencies
____ Quiz (7.3–7.4): SE page 429; CRB page 62

Notes ______________

LESSON 7.4

NAME ______________________________ DATE __________

Available as a transparency

Warm-Up Exercises

For use before Lesson 7.4, pages 421–430

Solve for *y* in each of the following.

1. $3y = 6x$
2. $2y = 4x + 2$
3. $3x + y = 6$
4. $2x + 6y = 6$
5. $3x + 8 = 4y$

Daily Homework Quiz

For use after Lesson 7.3, pages 415–420

Let $f(x) = x^2 - 2$ and $g(x) = x^3 + 4x$. Perform the indicated operation and state the domain.

1. $f(x) + g(x)$
2. $f(x) \cdot g(x)$
3. $f(g(x))$
4. $\dfrac{g(x)}{f(x)}$

Lesson 7.4

LESSON

7.4

NAME ______________________________ DATE __________

Activity Support Master

For use with page 421

Step 1 $y = f(x) = \dfrac{x - 3}{2}$

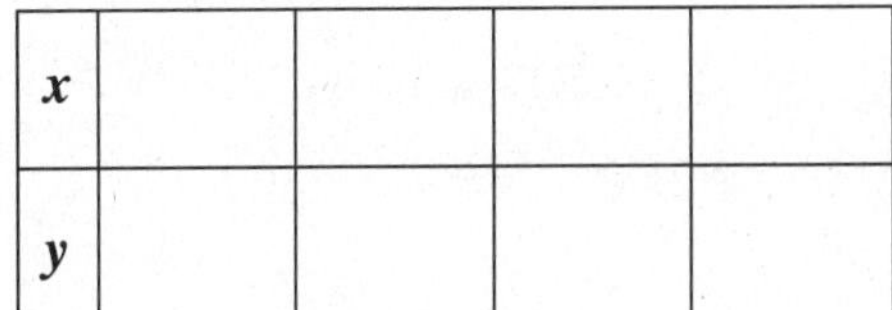

x				
y				

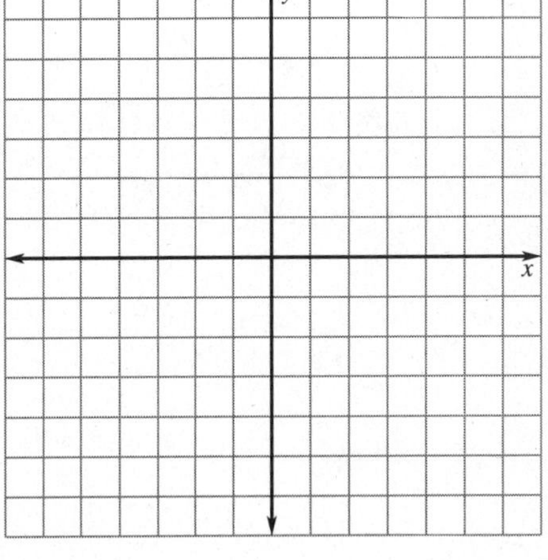

Step 2 Interchange x- and y- coordinates from **Step 1.**

x				
y				

Step 3

$g(x) =$ ______________________________

Step 4

____

____

____

Step 5

____

____

____

Step 6

____

____

____

Lesson 7.4

LESSON 7.4

NAME ______________________ DATE __________

Available as a transparency

Activity Lesson Opener

For use with pages 422–429

SET UP: Work individually.

Match each composition of functions on the left with a simplified expression on the right to reveal something you might say while riding a roller coaster. (A choice may be used more than once or not at all.)

1. $f(x) = 2x + 3$	_____	$f(g(x))$
$g(x) = 5x - 2$	_____	$g(f(x))$
2. $f(x) = x + 5$	_____	$f(g(x))$
$g(x) = x - 5$	_____	$g(f(x))$
3. $f(x) = 3x$	_____	$f(g(x))$
$g(x) = \frac{x}{3}$	_____	$g(f(x))$
4. $f(x) = 5x + 20$	_____	$f(g(x))$
$g(x) = \frac{1}{5}x - 4$	_____	$g(f(x))$
5. $f(x) = x^3 + 5$	_____	$f(g(x))$
$g(x) = \sqrt[3]{x - 5}$	_____	$g(f(x))$
6. $f(x) = \frac{x^5 - 8}{10}$	_____	$f(g(x))$
$g(x) = \sqrt[5]{10x + 8}$	_____	$g(f(x))$

A $9x$

B $x - 10$

C $x^5 - 8$

E x

G $9x$

H $10x + 13$

I $x - 5$

L x^5

O $x + 10$

P $x + 5$

R $\frac{x}{3}$

W $10x - 1$

U $3x$

Y $\sqrt[5]{x}$

7. Comment on any patterns you observe.

NAME ______________________________ DATE ____________

Graphing Calculator Activity Keystrokes

For use with page 430

TI-82

Y= 2 X,T,θ − 5

WINDOW ENTER (-) 15 ENTER 15 ENTER 1

ENTER (-) 10 ENTER 10 ENTER 1 ENTER

GRAPH

2nd [DRAW]8

2nd [Y-VARS]1 1 ENTER

TI-83

Y= 2 X,T,θ,n − 5

WINDOW (-) 15 ENTER 15 ENTER 1

ENTER (-) 10 ENTER 10 ENTER 1 ENTER

1 ENTER GRAPH

2nd [DRAW]8

VARS ▶ 1 1 ENTER

SHARP EL-9600c

Y= 2 X/θ/T/n − 5

WINDOW (-) 15 ENTER 15 ENTER 1

ENTER (-) 10 ENTER 10 ENTER 1 ENTER

GRAPH

2nd [DRAW][A]8

VARS [A] ENTER [A]1 ENTER

CASIO CFX-9850GA PLUS

From the main menu, choose GRAPH.

2 X,θ,T − 5 EXE

SHIFT F3 WINDOW (-) 15 EXE 15 EXE 1

EXE (-) 10 EXE 10 EXE 1 EXE EXIT F6

SHIFT F4 F4

LESSON 7.4

NAME ______________________ DATE __________

Practice A

For use with pages 422–429

Find the inverse relation.

1.

x	-2	-1	0	1	2
y	3	5	7	9	11

2.

x	0	1	2	3	4
y	1	-2	4	-1	0

Use the horizontal line test to determine whether the inverse of *f* is a function.

3.

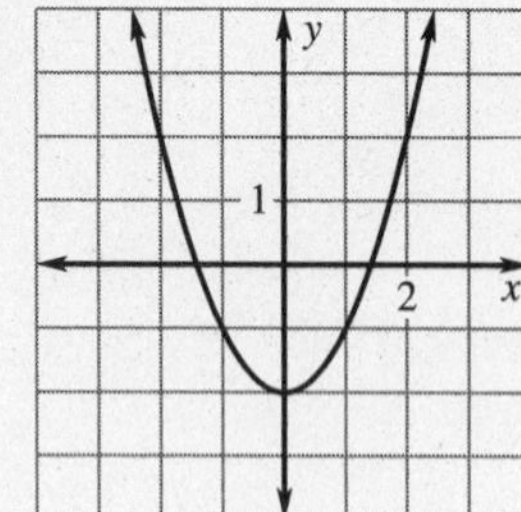

4.

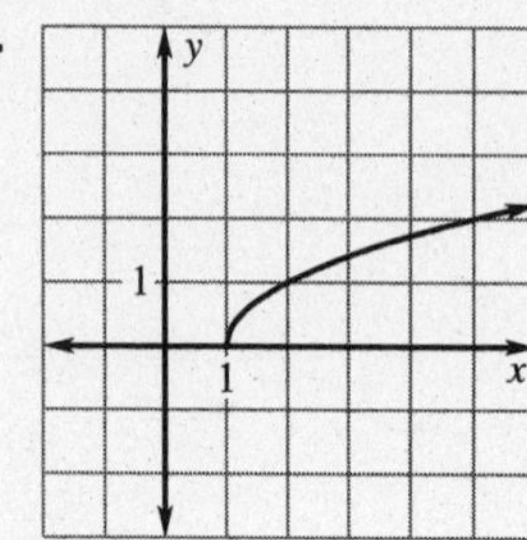

5.

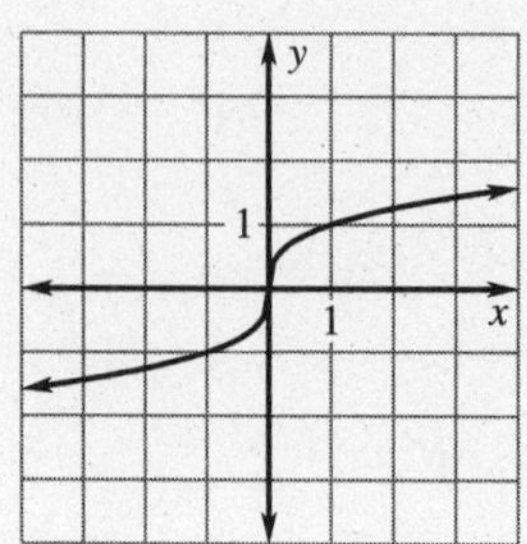

Match the graph with the graph of its inverse.

6.

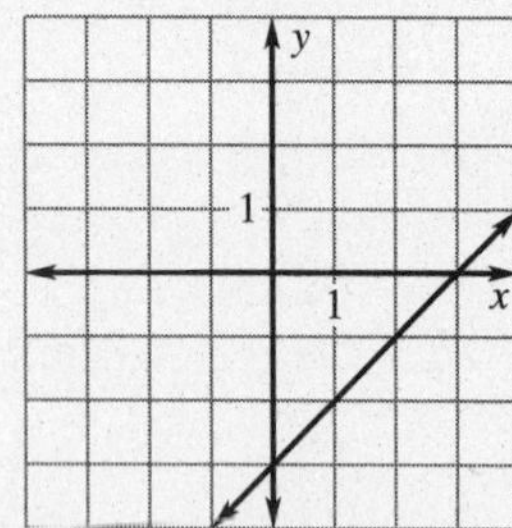

7.

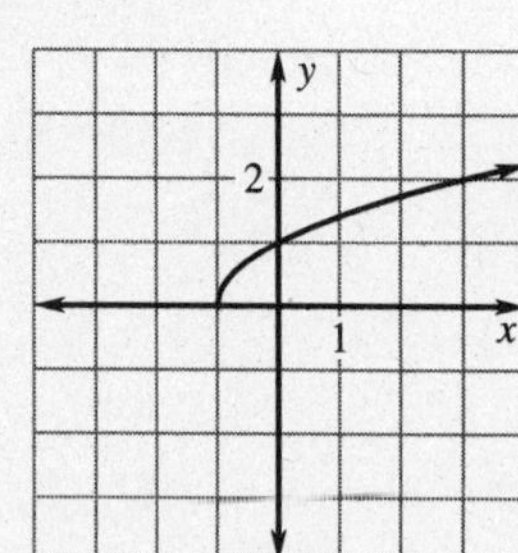

8.

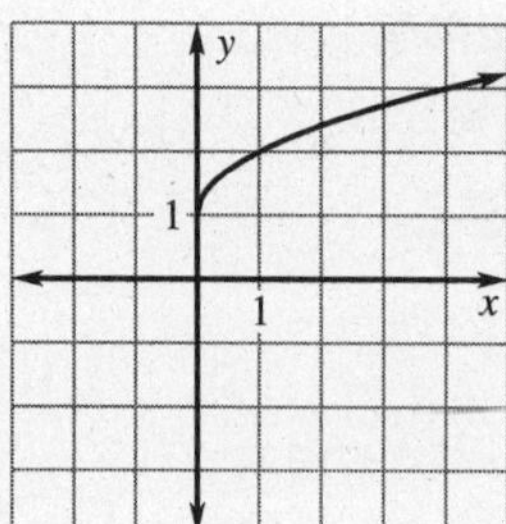

A.

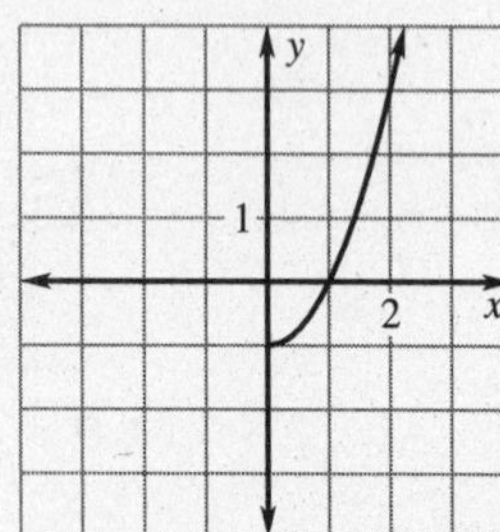

B.

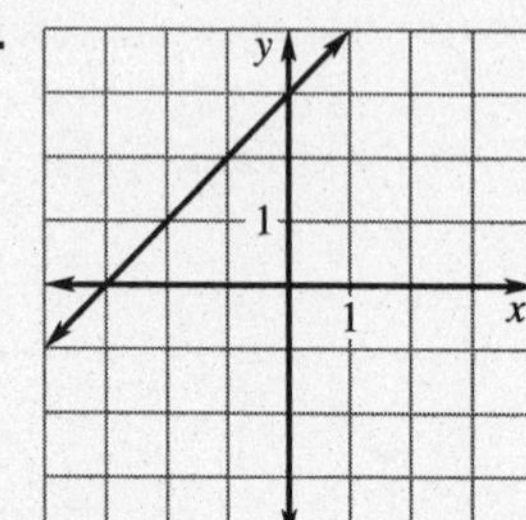

C.

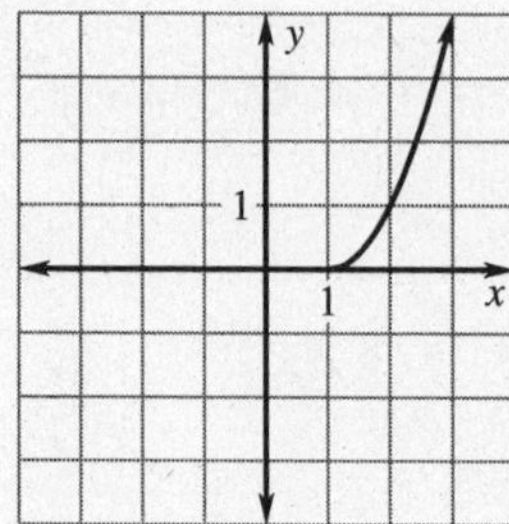

Verify that *f* and *g* are inverse functions.

9. $f(x) = x + 5,\ g(x) = x - 5$

10. $f(x) = 6x,\ g(x) = \frac{1}{6}x$

11. $f(x) = x^5,\ g(x) = \sqrt[5]{x}$

12. $f(x) = 2x + 1,\ g(x) = \frac{1}{2}x - \frac{1}{2}$

13. ***Metric Conversions*** The formula to convert inches to centimeters is $C = 2.54i$. Write the inverse function, which converts centimeters to inches. How many inches is 42 centimeters? Round your answer to two decimal places.

14. ***Geometry*** The formula $C = 2\pi r$ gives the circumference of a circle of radius r. Write the inverse function, which gives the radius of a circle of circumference C. What is the radius of a circle with a circumference of 28 inches? Round your answer to two decimal places.

LESSON 7.4

NAME ______________________ DATE __________

Practice B

For use with pages 422–429

Find the inverse of the relation.

1.

x	1	2	3	4	5
y	-6	-3	0	3	6

2.

x	-1	$-\frac{2}{3}$	0	$\frac{1}{2}$	3
y	1	2	4	6	0

Use the horizontal line test to determine whether the inverse of *f* is a function.

3. $f(x) = -3x + 5$
4. $f(x) = 2x^2 - 3$
5. $f(x) = 1 - x^2$
6. $f(x) = |x|$
7. $f(x) = 3 - 2x$
8. $f(x) = \frac{1}{2}x + 4$

Verify that *f* and *g* are inverse functions.

9. $f(x) = 2x,\ g(x) = \frac{x}{2}$
10. $f(x) = 1 - x,\ g(x) = 1 - x$
11. $f(x) = x - 2,\ g(x) = x + 2$
12. $f(x) = -3x + 6,\ g(x) = -\frac{1}{3}x + 2$
13. $f(x) = \frac{1}{2}x - 4,\ g(x) = 2x + 8$
14. $f(x) = 4x + 1,\ g(x) = \frac{1}{4}x - \frac{1}{4}$
15. $f(x) = x^2,\ x \geq 0,\ g(x) = \sqrt{x}$
16. $f(x) = x^3,\ g(x) = \sqrt[3]{x}$

Find an equation for the inverse of the relation.

17. $y = 4x$
18. $y = -x + 5$
19. $y = 3x + 1$
20. $y = 4x - 9$
21. $y = \frac{1}{2}x + 6$
22. $y = 3 - 2x$
23. $y = x^2 + 3$
24. $y = x^2 - 1$
25. $y = \sqrt{x}$

Sketch the inverse of *f* on the coordinate system.

26.

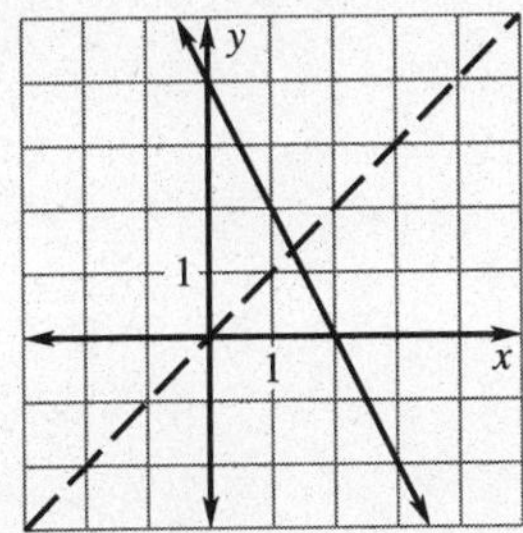

27.

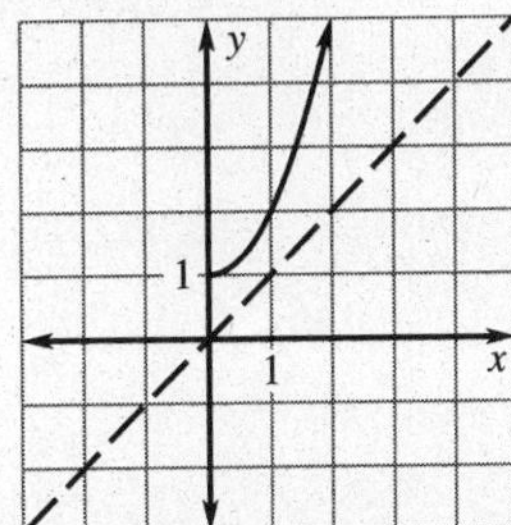

28.

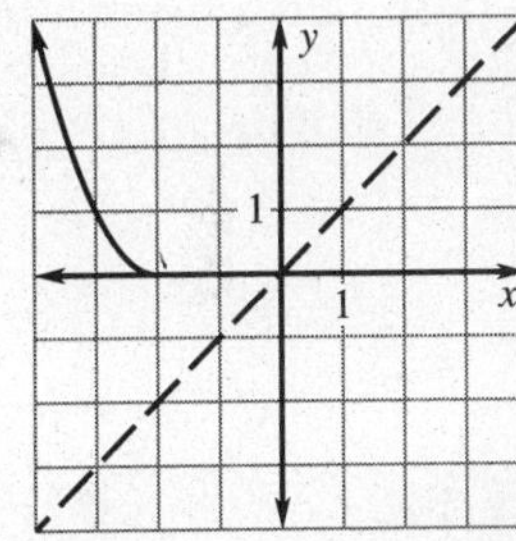

29. ***Temperature Conversion*** The formula to convert temperatures from degrees Celsius to Kelvins is $K = C + 273.15$. Write the inverse of the function, which converts temperatures from Kelvins to degrees Celsius. Then find the Celsius temperature that is equal to 295 Kelvins.

30. ***Sale Price*** A gift shop is having a storewide 25% off sale. The sale price S of an item that has a regular price of R is $S = R - 0.25R$. Write the inverse of the function. Then find the regular price of an item that you got for $19.88.

Lesson 7.4

NAME ______________________ DATE ________

Practice C

For use with pages 422–429

Verify that *f* and *g* are inverse functions.

1. $f(x) = 2x + 7,\ g(x) = \frac{1}{2}x - \frac{7}{2}$
2. $f(x) = -5x + 3,\ g(x) = \frac{3}{5} - \frac{1}{5}x$
3. $f(x) = \sqrt{x - 4},\ g(x) = x^2 + 4,\ x \geq 0$
4. $f(x) = \frac{1}{2}x^3,\ g(x) = \sqrt[3]{2x}$
5. $f(x) = \frac{1}{3}x^4 + 2,\ x \geq 0,\ g(x) = \sqrt[4]{3x - 6}$
6. $f(x) = \frac{3}{4}x^4 + 2,\ g(x) = \dfrac{\sqrt[4]{108x - 216}}{3}$

Find the inverse function.

7. $f(x) = 1 - 4x$
8. $f(x) = 3x + 8$
9. $f(x) = \sqrt{x + 1}$
10. $f(x) = \sqrt{2x - 3}$
11. $f(x) = \sqrt{4 - x}$
12. $f(x) = \sqrt[3]{5x - 3}$
13. $f(x) = x^2 + 7,\ x \geq 0$
14. $f(x) = 2x^3 + 5$
15. $f(x) = |x|,\ x \leq 0$

16. ***Critical Thinking*** Consider the basic power function $f(x) = x^n$ for $n = 1, 2, 3, 4,$ and 5. Make a conclusion about the values of n for which a restriction on the function's domain must be made to ensure that the inverse of f is a function.

17. ***Critical Thinking*** Consider the following pairs of inverse functions:

    $f(x) = 3x$ and $f^{-1}(x) = \frac{1}{3}x$

    $g(x) = \frac{2}{3}x$ and $g^{-1}(x) = \frac{3}{2}x$

    Does $f^{-1}(x) = \dfrac{1}{f(x)}$? Explain.

Visual Thinking **In Exercises 18–20, consider the function $f(x) = \dfrac{1}{x}$, which is its own inverse.**

18. Sketch the graph of $f(x)$ to verify that it is its own inverse.
19. Verify that $f(x)$ is its own inverse by showing $f(f(x)) = x$.
20. If $g(x) = af(x)$ where a is a nonzero constant, is it true that $g(x)$ is its own inverse? Explain.

Use the horizontal line test to determine whether the inverse of the function is a function.

21. $f(x) = \begin{cases} x + 2, & x < 0 \\ x + 1, & x \geq 0 \end{cases}$

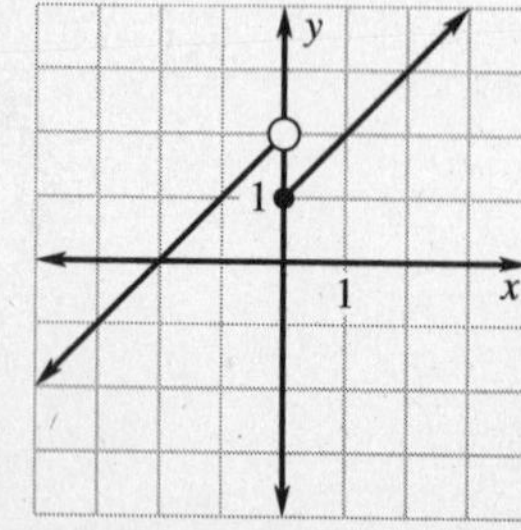

22. $f(x) = \begin{cases} x + 2, & x < 0 \\ x + 3, & x \geq 0 \end{cases}$

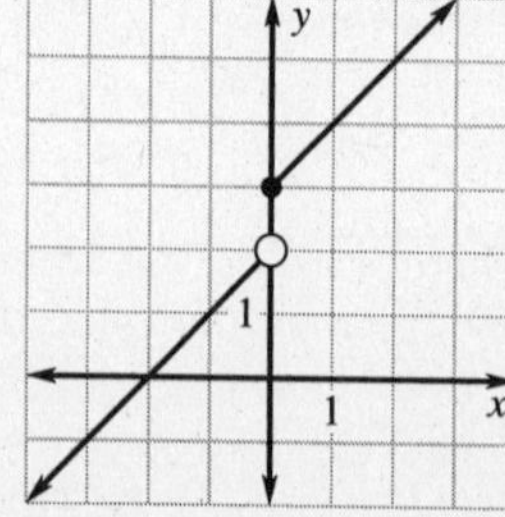

23. $f(x) = \begin{cases} x^2, & x < 0 \\ -x, & x \geq 0 \end{cases}$

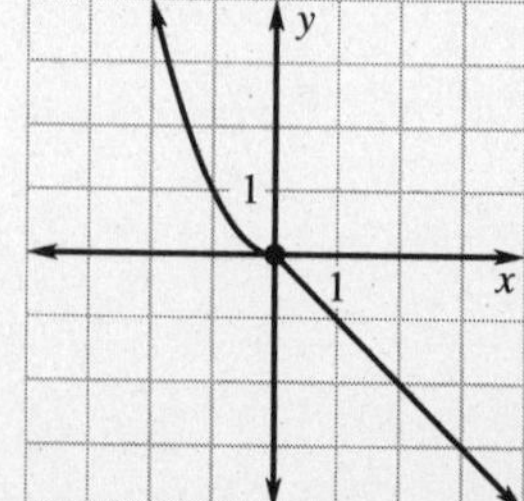

LESSON 7.4

NAME ______________________ DATE __________

Reteaching with Practice

For use with pages 422–429

GOAL Find inverses of linear and nonlinear functions

VOCABULARY

An **inverse relation** maps the output values back to their original input values. Two functions f and g are called **inverse functions** provided $f(g(x)) = x$ and $g(f(x)) = x$. According to the **horizontal line test,** if no horizontal line intersects the graph of a function f more than once, then the inverse of f is itself a function.

EXAMPLE 1 Finding an Inverse Relation

Find an equation for the inverse of the relation $y = \frac{1}{3}x + 2$.

SOLUTION

$y = \frac{1}{3}x + 2$	Write original relation.
$x = \frac{1}{3}y + 2$	Switch x and y.
$x - 2 = \frac{1}{3}y$	Subtract 2 from each side.
$3x - 6 = y$	Multiply each side by 3, the reciprocal of $\frac{1}{3}$.

The inverse relation is $y = 3x - 6$.

Exercises for Example 1

Find an equation for the inverse relation.

1. $y = 4x + 8$

2. $y = -3x + 12$

3. $y = \frac{2}{3}x - 4$

EXAMPLE 2 Verifying Inverse Functions

Verify that $f(x) = \frac{1}{2}x - 2$ and $f^{-1}(x) = 2x + 4$ are inverses.

SOLUTION

You need to show that $f(f^{-1}(x)) = x$ and $f^{-1}(f(x)) = x$.

$$\begin{aligned} f(f^{-1}(x)) &= f(2x + 4) \\ &= \tfrac{1}{2}(2x + 4) - 2 \\ &= x + 2 - 2 \\ &= x \checkmark \end{aligned}$$

$$\begin{aligned} f^{-1}(f(x)) &= f^{-1}\left(\tfrac{1}{2}x - 2\right) \\ &= 2\left(\tfrac{1}{2}x - 2\right) + 4 \\ &= x - 4 + 4 \\ &= x \checkmark \end{aligned}$$

Exercises for Example 2

Verify that f and g are inverse functions.

4. $f(x) = x - 3$, $g(x) = x + 3$

5. $f(x) = \frac{1}{2}x + 3$, $g(x) = 2x - 6$

NAME ______________________________ DATE __________

Reteaching with Practice

For use with pages 422–429

EXAMPLE 3 Finding an Inverse Power Function

Find the inverse of the function $f(x) = x^5$

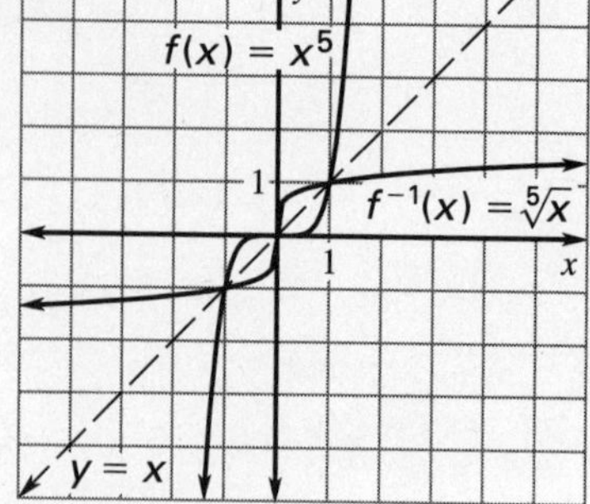

SOLUTION

$f(x) = x^5$ Write original function.

$y = x^5$ Replace $f(x)$ with y.

$x = y^5$ Switch x and y.

$\sqrt[5]{x} = y$ Take fifth roots of each side.

The inverse function is $f^{-1}(x) = \sqrt[5]{x}$.

To check your work, you could graph f and f^{-1} on the same coordinate axes. Notice that the graph of f^{-1} is the reflection of f in the line $y = x$.

Exercises for Example 3

Find the inverse power function.

6. $f(x) = -x^2, x \geq 0$ **7.** $f(x) = -27x^3$ **8.** $f(x) = x^4, x \leq 0$

EXAMPLE 4 Finding an Inverse Function

Consider the function $f(x) = -x^2 + 5$. Determine whether the inverse of f is a function.

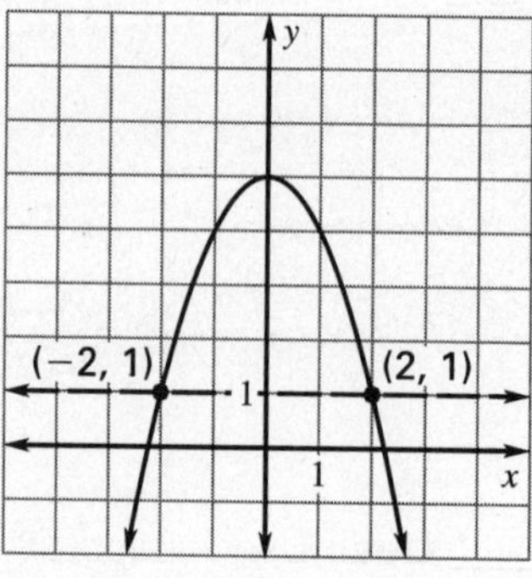

SOLUTION

Begin by graphing the function. The function is a parabola which opens down and is vertically shifted five units up. Notice that a horizontal line, such as $y = 1$, intersects the graph more than once. This tells you that the inverse of f is not a function.

Exercises for Example 4

Graph the function f. Then use the graph to determine whether the inverse of f is a function. If it is, find the inverse.

9. $f(x) = 3x - 4$ **10.** $f(x) = x^4 + 2$ **11.** $f(x) = x^3 - 4$

12. $f(x) = 5x^2$ **13.** $f(x) = -x^3$ **14.** $f(x) = |x|$

LESSON 7.4

NAME ______________________________ DATE ________

Quick Catch-Up for Absent Students

For use with pages 421–430

The items checked below were covered in class on (date missed) ____________

Activity 7.4: Exploring Inverse Functions (p. 421)

____ **Goal:** Find how a function and its inverse are related.

Lesson 7.4: Inverse Functions

____ **Goal 1:** Find inverses of linear functions. (p. 422, 423)

Material Covered:

____ Student Help: Look Back

____ Example 1: Finding an Inverse Relation

____ Student Help: Study Tip

____ Example 2: Verifying Inverse Functions

____ Student Help: Study Tip

____ Example 3: Writing an Inverse Model

Vocabulary:

inverse relation, p. 422 inverse funcion, p. 422

____ **Goal 2:** Find inverses of nonlinear functions. (p. 424, 425)

Material Covered:

____ Example 4: Finding an Inverse Power Function

____ Example 5: Finding an Inverse Function

____ Example 6: Modeling with an Inverse Function

Activity 7.4: Graphing Inverse Functions (p. 430)

____ **Goal:** Use a graphing calculator to graph inverse functions.

____ Student Help: Keystroke Help

____ Other (specify) ____

____

Homework and Additional Learning Support

____ Textbook (specify) pp. 426–429

____

____ *Reteaching with Practice* worksheet (specify exercises) ______________

____ *Personal Student Tutor* for Lesson 7.4

Lesson 7.4

LESSON 7.4

NAME ______________________ DATE __________

Real-Life Application: When Will I Ever Use This?

For use with pages 422–429

Slam Dunk

A dunk is defined as a field goal that is made by a basketball player slamming the basketball through the hoop. A player may use use both hands or hold the ball in one hand. Sometimes another player will pass the ball to a second player who catches the ball while in mid-air and slams it through the hoop.

Every year the NBA holds an All-Star game. During the weekend of the All-Star game, other competitions are put on by the NBA, such as the slam dunk contest. Six players are invited to participate in the slam dunk contest based on their performance throughout the season. The six players are judged on "creativity, artistry, and athletic ability" displayed in their dunking skills. Three players move on to a second round where one will be named the slam dunk champion.

In Exercises 1–3, use the following information.

A basketball hoop is 10 feet above the floor. A player must jump high enough to reach $10\frac{1}{2}$ feet to make a slam dunk. A person's height H in feet can be modeled by

$$H = \frac{2}{3}R + 1$$

where R is the person's standing reach in feet.

1. Write the inverse of the function.

2. Find the standing reach of a person with a height of 6 feet.

3. Find the standing reach of a person with a height of 6 feet and 6 inches.

In Exercises 4 and 5, use the following information.

The initial velocity, v (in feet per second), at which a person must jump to make a slam dunk is given by $R = \frac{21}{2} - \frac{1}{64}v^2$.

4. Write the inverse of the function.

5. Copy and complete the table.

Height, H	5.5	6.0	6.5	7.0
Reach, R				
Velocity, v				

Lesson 7.4

LESSON **7.4**

NAME ____________________ DATE ________

Challenge: Skills and Applications

For use with pages 422–429

1. a. Find an expression for the inverse function for the general linear function $f(x) = mx + b$, with $m \neq 0$, in terms of m, b, and x.

b. Find, in terms of m and b, the coordinates of the point where the graph of the function and the graph of the inverse function intersect.

c. If F is a temperature expressed in degrees Fahrenheit, and C is the same temperature expressed in degrees Celsius, $F = \frac{9}{5}C + 32$. Use your answer to part (b) to find the temperature that is the same on both temperature scales.

2. The graphs of a function $f(x)$ and its inverse function $f^{-1}(x)$ are shown along with a tangent line to each graph at corresponding points.

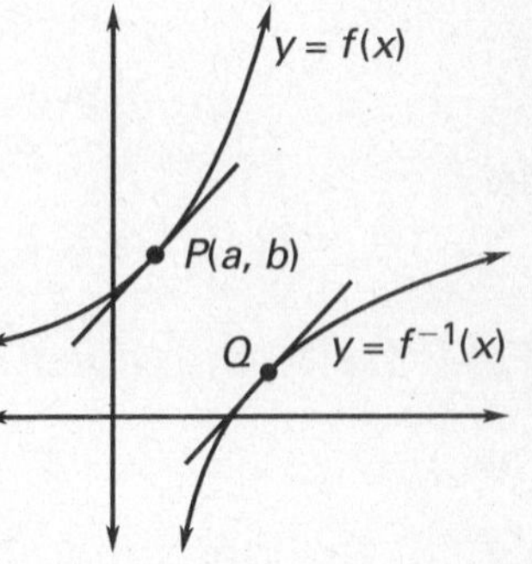

a. What are the coordinates of the point Q on the graph of $f^{-1}(x)$?

b. What appears to be the relationship between the linear functions that define the tangent lines at the two points?

c. Use your answer to part (a) of Exercise 1 to state a general relationship between the slopes of the tangent lines at a point on the graph of the function and at a point on the graph of the inverse function.

For Exercises 3–4, find the inverse function of each function.

3. $f(x) = \begin{cases} x + 1 \text{ if } x \geq 0 \\ 2x + 1 \text{ if } x < 0 \end{cases}$

4. $f(x) = x|x| = \begin{cases} x^2 \text{ if } x \geq 0 \\ -x^2 \text{ if } x < 0 \end{cases}$

5. Let a be a real number between 0 and 1 (not including 1), so that a has a decimal expansion of the form

$$a = 0.a_1a_2a_3\ldots,$$

where each a_j is one of the digits 0, 1, 2, . . . , 9. Let $f(x)$ and $g(x)$ be functions whose domain consists of all such numbers and are defined as follows:

$$f(a) = 0.a_2a_3a_4\ldots \qquad g(a) = 0.0a_1a_2a_3\ldots$$

Find $f(g(a))$ and $g(f(a))$. Are f and g inverse functions?

Lesson 7.4

NAME ______________________________ DATE ____________

Quiz 2

For use after Lessons 7.3–7.4

Let $f(x) = 5x^2 - x^{\frac{1}{3}}$ and $g(x) = 4x^{\frac{1}{3}}$. Perform the indicated operation and state the domain. *(Lesson 7.3)*

1. $f(x) - g(x)$
2. $f(x) \cdot g(x)$
3. $f(x) + g(x)$
4. $\dfrac{f(x)}{g(x)}$

Let $f(x) = 4x^{-1}$ and $g(x) = x - 6$. Perform the indicated operation and state the domain. *(Lesson 7.3)*

5. $f(g(x))$
6. $g(f(x))$
7. $f(f(x))$

Verify that f and g are inverse functions. *(Lesson 7.4)*

8. $f(x) = 3x - 2,\ g(x) = \dfrac{x+2}{3}$
9. $f(x) = (x - 2)^{\frac{1}{4}},\ g(x) = x^4 + 2$

Find the inverse function. *(Lesson 7.4)*

10. $f(x) = 2x + 7$
11. $g(x) = 3x^4,\ x > 0$

Graph the function f. Then use the graph to determine whether the inverse of f is a function. *(Lesson 7.4)*

12. $f(x) = 5x^6 + 3$
13. $f(x) = \dfrac{1}{2}(x - 1)^3$

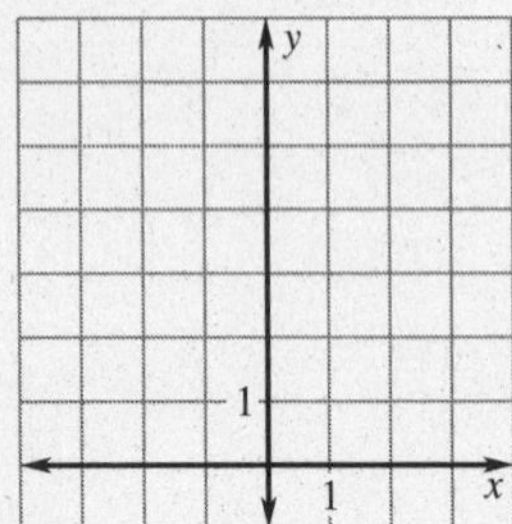

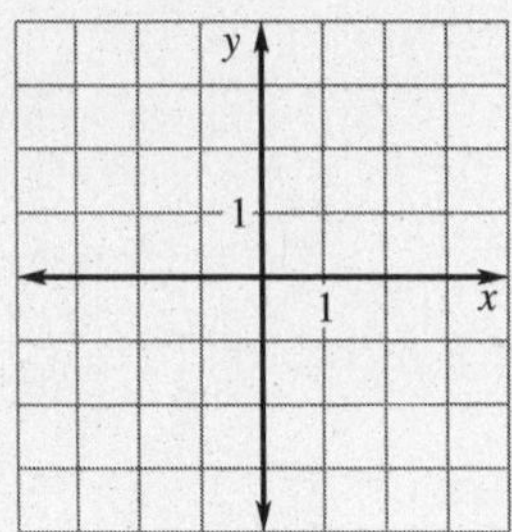

Answers

1. ____________
2. ____________
3. ____________
4. ____________
5. ____________
6. ____________
7. ____________
8. ____________
9. ____________
10. ____________
11. ____________
12. Use grid at left.
13. Use grid at left.

LESSON 7.5

TEACHER'S NAME ____________ CLASS ______ ROOM ______ DATE ______

Lesson Plan

1-day lesson (See *Pacing the Chapter,* TE pages 398C–398D) For use with pages 431–436

GOALS
1. Graph square root and cube root functions.
2. Use square root and cube root functions to find real-life quantities.

State/Local Objectives ____________

____

✓ Check the items you wish to use for this lesson.

STARTING OPTIONS

____ Homework Check: TE page 426; Answer Transparencies
____ Warm-Up or Daily Homework Quiz: TE pages 431 and 429, CRB page 65, or Transparencies

TEACHING OPTIONS

____ Lesson Opener (Visual Approach): CRB page 66 or Transparencies
____ Graphing Calculator Activity with Keystrokes: CRB page 67
____ Examples 1–6: SE pages 432–433
____ Extra Examples: TE pages 432–433 or Transparencies
____ Closure Question: TE page 433
____ Guided Practice Exercises: SE page 434

APPLY/HOMEWORK

Homework Assignment

____ Basic 15–21, 22–40 even, 46, 50, 55–69 odd
____ Average 15–21, 22–50 even, 55–69 odd
____ Advanced 15–21, 22–50 even, 51–53, 55–69 odd, 70

Reteaching the Lesson

____ Practice Masters: CRB pages 68–70 (Level A, Level B, Level C)
____ Reteaching with Practice: CRB pages 71–72 or Practice Workbook with Examples
____ Personal Student Tutor

Extending the Lesson

____ Applications (Interdisciplinary): CRB page 74
____ Challenge: SE page 436; CRB page 75 or Internet

ASSESSMENT OPTIONS

____ Checkpoint Exercises: TE pages 432–433 or Transparencies
____ Daily Homework Quiz (7.5): TE page 436, CRB page 78, or Transparencies
____ Standardized Test Practice: SE page 436; TE page 436; STP Workbook; Transparencies

Notes ____________

____

____

Lesson 7.5

LESSON 7.5

TEACHER'S NAME ______________________ CLASS ______ ROOM ______ DATE ______

Lesson Plan for Block Scheduling

Half-day lesson (See *Pacing the Chapter,* TE pages 398C–398D) **For use with pages 431–436**

GOALS
1. **Graph square root and cube root functions.**
2. **Use square root and cube root functions to find real-life quantities.**

State/Local Objectives ____

____

____

CHAPTER PACING GUIDE	
Day	**Lesson**
1	Assess Ch. 6; 7.1(all)
2	7.2 (all)
3	7.3 (all); 7.4(begin)
4	7.4 (end); **7.5(all)**
5	7.6 (all)
6	7.7 (all)
7	Review/Assess Ch. 7

✓ Check the items you wish to use for this lesson.

STARTING OPTIONS
____ Homework Check: TE page 426; Answer Transparencies
____ Warm-Up or Daily Homework Quiz: TE pages 431 and 429, CRB page 65, or Transparencies

TEACHING OPTIONS
____ Lesson Opener (Visual Approach): CRB page 66 or Transparencies
____ Graphing Calculator Activity with Keystrokes: CRB page 67
____ Examples 1–6: SE pages 432–433
____ Extra Examples: TE pages 432–433 or Transparencies
____ Closure Question: TE page 433
____ Guided Practice Exercises: SE page 434

APPLY/HOMEWORK
Homework Assignment (See also the assignment for Lesson 7.4.)
____ Block Schedule: 15–21, 22–50 even, 55–69 odd

Reteaching the Lesson
____ Practice Masters: CRB pages 68–70 (Level A, Level B, Level C)
____ Reteaching with Practice: CRB pages 71–72 or Practice Workbook with Examples
____ Personal Student Tutor

Extending the Lesson
____ Applications (Interdisciplinary): CRB page 74
____ Challenge: SE page 436; CRB page 75 or Internet

ASSESSMENT OPTIONS
____ Checkpoint Exercises: TE pages 432–433 or Transparencies
____ Daily Homework Quiz (7.5): TE page 436, CRB page 78, or Transparencies
____ Standardized Test Practice: SE page 436; TE page 436; STP Workbook; Transparencies

Notes ____

____

____

LESSON 7.5

NAME ______________________ DATE ________

Available as a transparency

WARM-UP EXERCISES

For use before Lesson 7.5, pages 431–436

1. Graph $y = -2x^2 + 3$.

2. Graph $y = x^3 - 2$.

DAILY HOMEWORK QUIZ

For use after Lesson 7.4, pages 421–430

Find the inverse relation.

1.

x	-2	-1	0	1	2
y	4	2	0	2	4

2. $y = 3x - 4$

Find the inverse function.

3. $f(x) = 2x^5$

4. $f(x) = 4x^3 - 1$

LESSON 7.5

NAME ______________________ DATE __________

Available as a transparency

Visual Approach Lesson Opener

For use with pages 431–436

When you studied quadratic functions, you learned that the graph of $y = (x - h)^2 + k$ has its vertex at (h, k).

The graph of $y = (x - 2)^2 - 3$ is the same as the graph of $y = x^2$, shifted 2 units to the right and 3 units down.

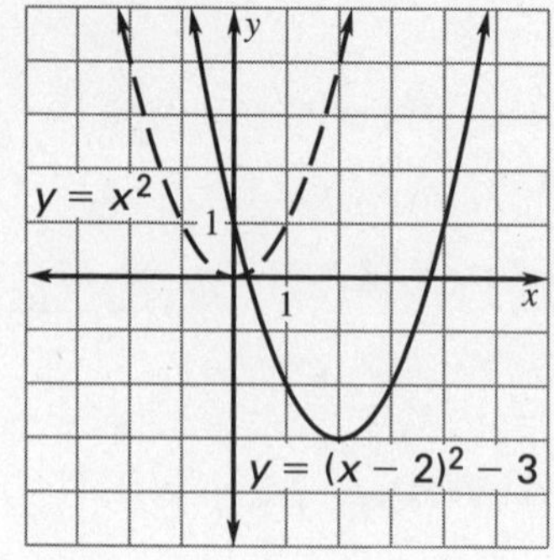

Similarly, the graph of $y = \sqrt{x - 2} - 3$ is the same as the graph of $y = \sqrt{x}$, shifted 2 units to the right and 3 units down.

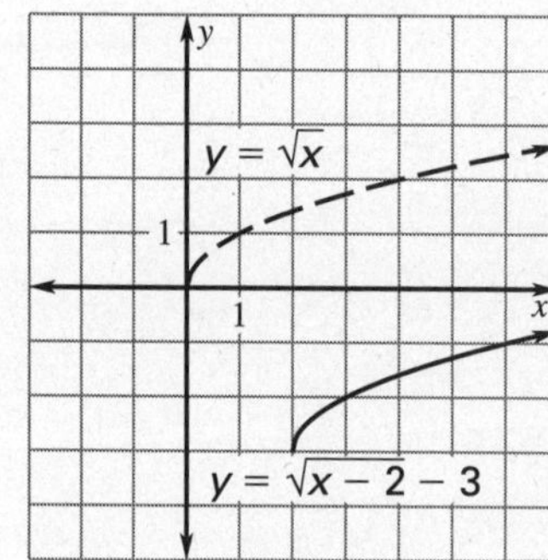

In general, the graph of $y = \sqrt{x - h} + k$ is the same as the graph of $y = \sqrt{x}$, shifted h units to the right and k units up.

Graph each function.

1. $y = \sqrt{x - 1} + 2$

2. $y = \sqrt{x - 3} - 2$

3. $y = \sqrt{x + 4} - 3$

4. $y = \sqrt{x + 2} + 1$

NAME ______________________ DATE ____________

Graphing Calculator Activity Keystrokes

For use with page 433

Keystrokes for Example 6

TI-82

Y= 62.5 MATH 4 X,T,θ + 75.8 ENTER 200 ENTER

WINDOW ENTER 0 ENTER 15 ENTER 1 ENTER 100 ENTER 250 ENTER 10 ENTER

GRAPH

Find point of intersection near $x \approx 7.85$.

2nd [CALC]5 ENTER ENTER

Use cursor to select guess at $x \approx 8$.

ENTER

TI-83

Y= 62.5 MATH 4 X,T,θ,n + 75.8 ENTER 200 ENTER

WINDOW 0 ENTER 15 ENTER 1 ENTER 100 ENTER 250 ENTER 10 ENTER

GRAPH

Find point of intersection near $x \approx 7.85$.

2nd [CALC]5 ENTER ENTER 7.85 ENTER

SHARP EL-9600c

Y= 62.5 (3 2ndF [$a\sqrt{\ }$] X/θ/T/n ▶)

+ 75.8 ENTER 200 ENTER

WINDOW 0 ENTER 15 ENTER 1 ENTER

100 ENTER 250 ENTER 10 ENTER GRAPH

Find point of intersection near $x \approx 7.85$.

2ndF [CALC]2

CASIO CFX-9850GA PLUS

From the main menu, choose GRAPH.

62.5 SHIFT [$\sqrt[3]{\ }$] X,T,θ + 75.8 EXE 200

EXE SHIFT 0 EXE 15 EXE 1 EXE 100 EXE

250 EXE 10 EXE EXIT F6

Find point of intersection near $x \approx 7.85$.

SHIFT F5 F5

LESSON 7.5

NAME ______________________ DATE ________

Practice A

For use with pages 431–436

Match the function with its graph.

1. $f(x) = \sqrt[3]{x} + 2$
2. $f(x) = \sqrt[3]{x} - 2$
3. $f(x) = \sqrt[3]{x - 2}$
4. $f(x) = \sqrt{x + 1}$
5. $f(x) = -\sqrt{x + 1}$
6. $f(x) = \sqrt{x - 1}$

A.
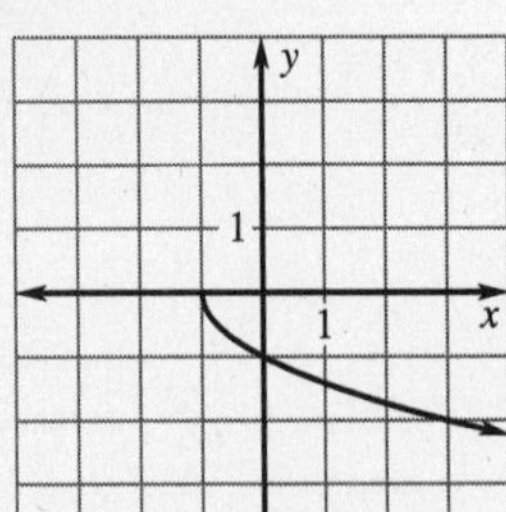

B.
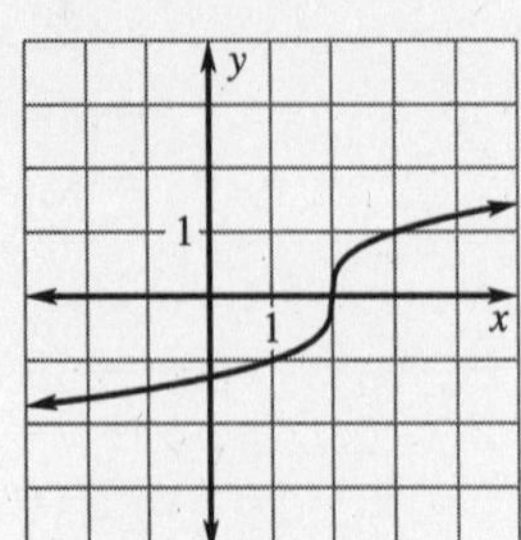

C.
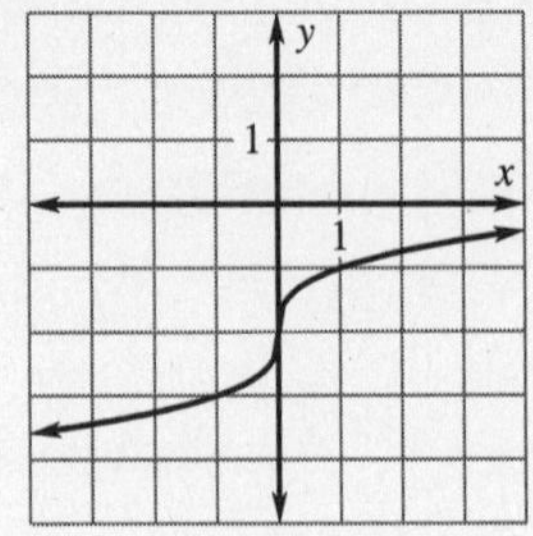

D.
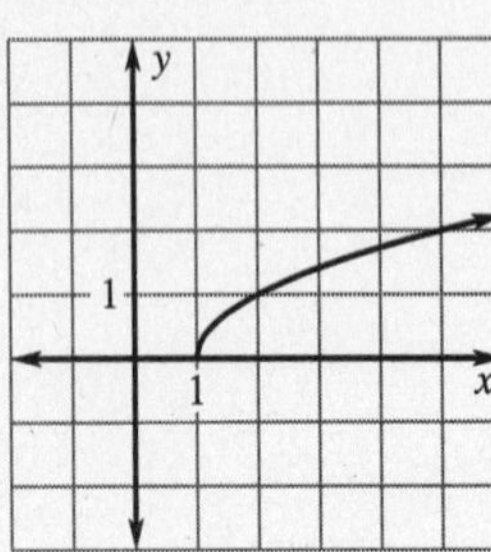

E.
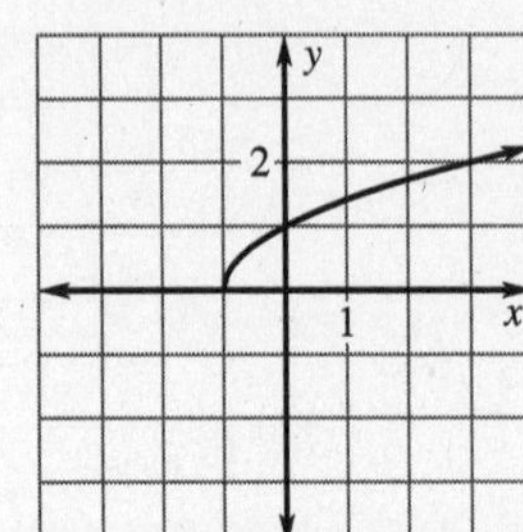

F.
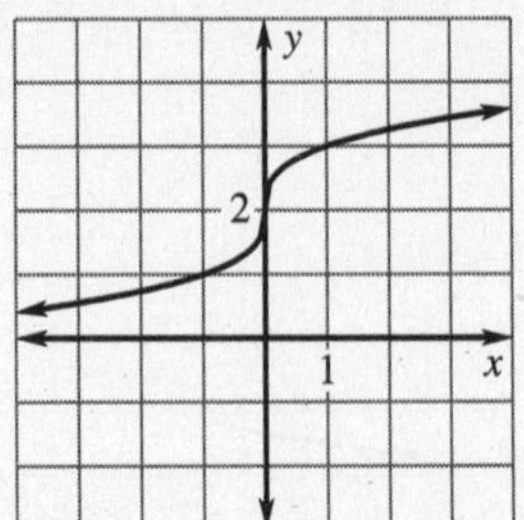

Describe how to obtain the graph of *g* from the graph of $f(x) = \sqrt{x}$.

7. $g(x) = \sqrt{x} + 3$
8. $g(x) = \sqrt{x} - 2$
9. $g(x) = -\sqrt{x}$
10. $g(x) = \sqrt{x + 1}$
11. $g(x) = \sqrt{x - 4}$
12. $g(x) = 2\sqrt{x}$

Describe how to obtain the graph of *g* from the graph of $f(x) = \sqrt[3]{x}$.

13. $g(x) = \sqrt[3]{x} - 3$
14. $g(x) = \sqrt[3]{x} + 2$
15. $g(x) = \sqrt[3]{x + 7}$
16. $g(x) = \sqrt[3]{x - 5}$
17. $g(x) = \frac{1}{2}\sqrt[3]{x}$
18. $g(x) = -\sqrt[3]{x}$

Graph the function. Then state the domain and range.

19. $f(x) = \sqrt{x} + 4$
20. $f(x) = \sqrt{x} - 3$
21. $f(x) = \sqrt{x + 2}$
22. $f(x) = \sqrt{x - 3}$
23. $f(x) = \sqrt[3]{x} + 1$
24. $f(x) = \sqrt[3]{x + 2}$

Falling Object **In Exercises 25–27, use the following information.**

A stone is dropped from a height of 100 feet. The time it takes for the stone to reach a height of h feet is given by the function $t = \frac{1}{4}\sqrt{100 - h}$ where t is time in seconds.

25. Identify the domain and range of the function.
26. Sketch the graph of the function.
27. What is the height of the stone after 2 seconds?

NAME _______________________________________ DATE __________

Practice B

For use with pages 431–436

Match the function with its graph.

1. $f(x) = \sqrt{x - 1} - 1$ **2.** $f(x) = \sqrt{x - 1} + 1$ **3.** $f(x) = \sqrt{x + 1} - 1$

4. $f(x) = \sqrt[3]{x + 2} - 1$ **5.** $f(x) = \sqrt[3]{x + 2} + 1$ **6.** $f(x) = \sqrt[3]{x - 2} - 1$

A.

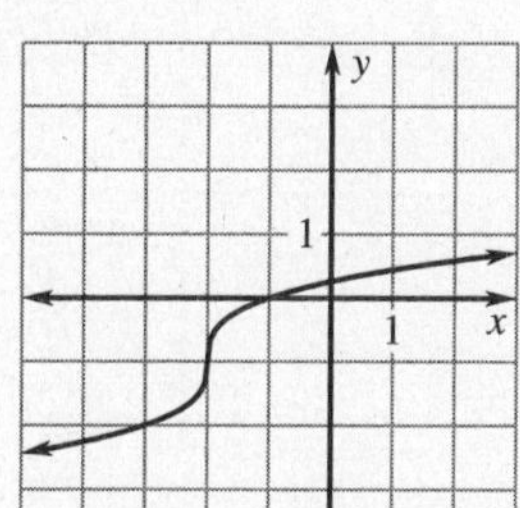

B.

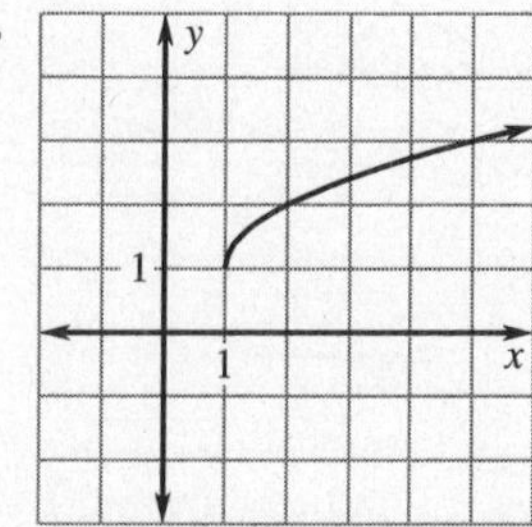

C.

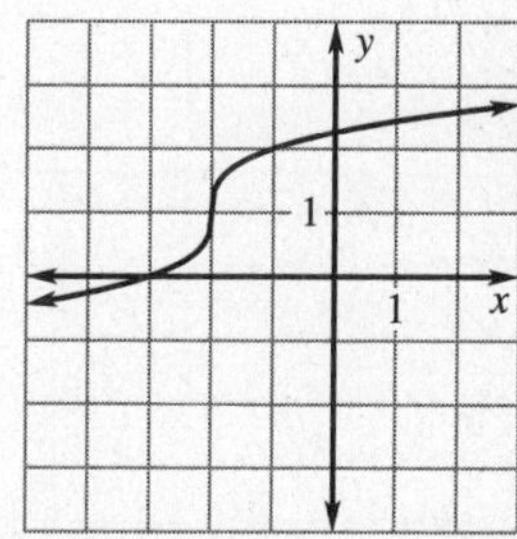

D.

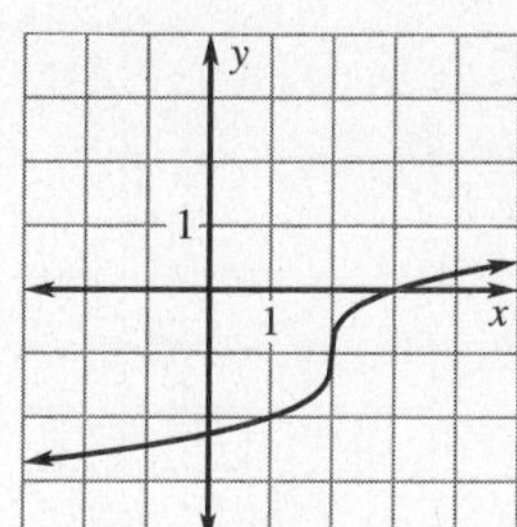

E.

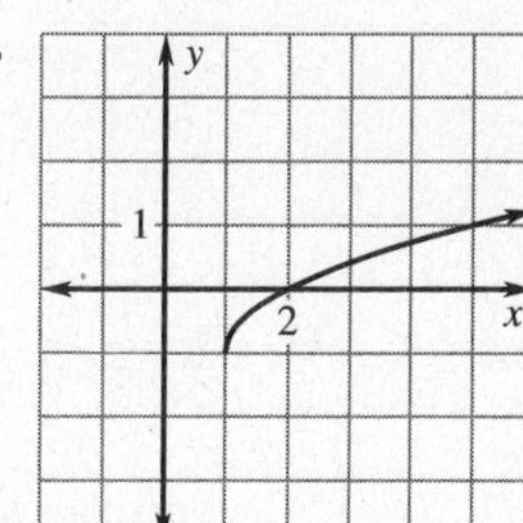

F.

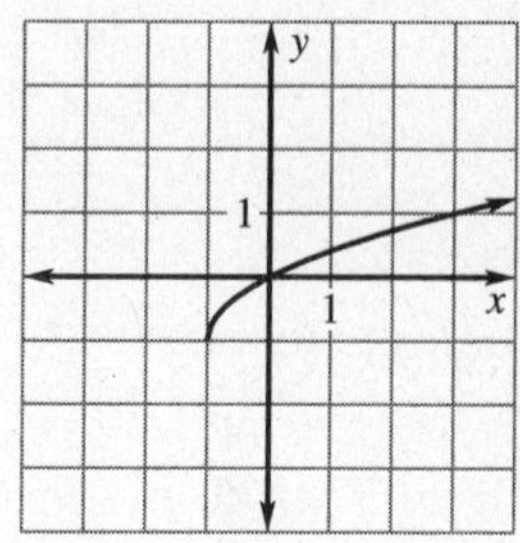

Describe how to obtain the graph of *g* from the graph of $f(x) = \sqrt{x}$.

7. $g(x) = \sqrt{x + 4} + 3$ **8.** $g(x) = \sqrt{x + 4} - 2$ **9.** $g(x) = -\sqrt{x + 4}$

10. $g(x) = \sqrt{x - 4} - 3$ **11.** $g(x) = \sqrt{x - 4} + 2$ **12.** $g(x) = -\sqrt{x - 4} + 2$

Describe how to obtain the graph of *g* from the graph of $f(x) = \sqrt[3]{x}$.

13. $g(x) = -\sqrt[3]{x} - 1$ **14.** $g(x) = -\sqrt[3]{x} + 1$ **15.** $g(x) = \sqrt[3]{x - 1} + 5$

16. $g(x) = \sqrt[3]{x + 1} + 5$ **17.** $g(x) = \sqrt[3]{x + 1} - 2$ **18.** $g(x) = \sqrt[3]{x - 1} - 2$

Graph the function. Then state the domain and range.

19. $f(x) = \sqrt{x - 3} + 2$ **20.** $f(x) = \sqrt{x + 1} - 3$ **21.** $f(x) = -\sqrt{x + 1} + 3$

22. $f(x) = \sqrt[3]{x + 1} + 3$ **23.** $f(x) = \sqrt[3]{x - 4} - 2$ **24.** $f(x) = -\sqrt[3]{x + 1} - 3$

Speed of Sound **In Exercises 25–27, use the following information.**

The speed of sound in feet per second through air of any temperature measured in Celsius is given by

$$V = \frac{1087\sqrt{273 + t}}{16.52},$$

where t is the temperature.

25. Identify the domain and range of the function.

26. Sketch the graph of the function.

27. What is the temperature of the air if the speed of sound is 1110 feet per second?

NAME ______________________________ DATE ___________

Practice C

For use with pages 431–436

Match the function with its graph.

1. $f(x) = \sqrt{x+2} - 1$
2. $f(x) = -\sqrt{x+2} - 1$
3. $f(x) = -\sqrt{x+2} + 1$

A.

B. 

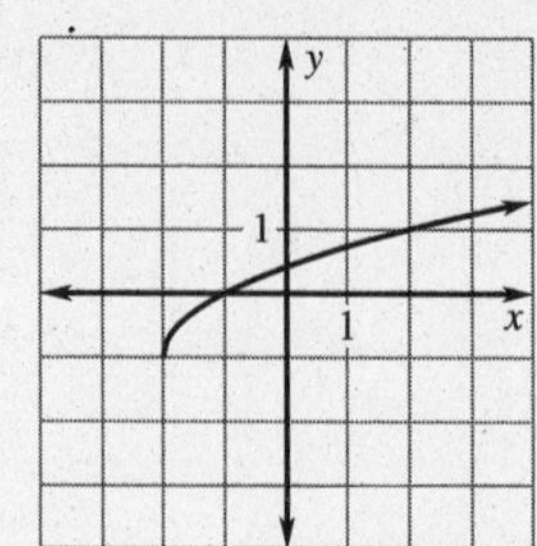

C.

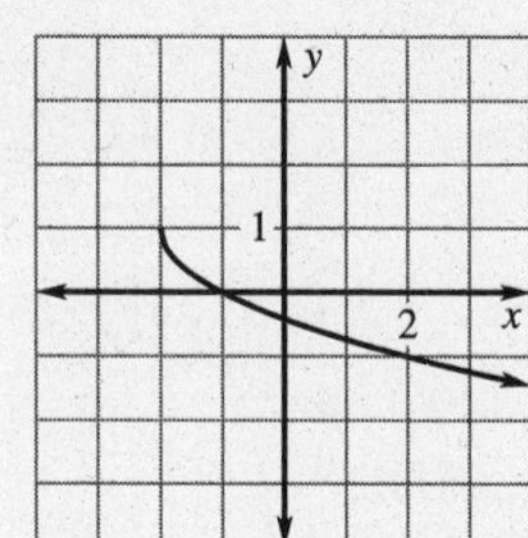

Sketch the graph of the function. Then state the domain and range.

4. $f(x) = 2\sqrt{x-3}$
5. $f(x) = \frac{1}{2}\sqrt{x+4}$
6. $f(x) = \sqrt{x+1} - 4$
7. $f(x) = -\frac{2}{3}\sqrt{x-1} + 3$
8. $f(x) = -\sqrt{x+1} - 2$
9. $f(x) = \sqrt{x+\frac{1}{2}} - 2$
10. $f(x) = 4\sqrt[3]{x+1}$
11. $f(x) = \frac{3}{2}\sqrt[3]{x-2}$
12. $f(x) = \sqrt[3]{x-3} + 2$
13. $f(x) = -\sqrt[3]{x-3} + 1$
14. $f(x) = -\frac{4}{5}\sqrt[3]{x+1} - 2$
15. $f(x) = \sqrt[3]{x-1} - \frac{1}{3}$

Visual Thinking **In Exercises 16–18, use the following information.**

Graph the functions $f(x) = \sqrt{x}$, $g(x) = \sqrt[4]{x}$, $h(x) = \sqrt[6]{x}$, and $j(x) = \sqrt[8]{x}$ on the same coordinate plane. Use the window xmin $= -1$, xmax $= 2$, xscl $= 1$, ymin $= -1$, ymax $= 2$, and Yscl $= 1$.

16. What two points do all of the graphs have in common?
17. Describe how the graphs are related.
18. Using what you have learned in Exercises 16 and 17, sketch the graph of $f(x) = \sqrt[4]{x-3} + 2$.

Visual Thinking Graph the functions $f(x) = \sqrt[3]{x}$, $g(x) = \sqrt[5]{x}$, $h(x) = \sqrt[7]{x}$, and $j(x) = \sqrt[9]{x}$ on the same coordinate plane. Use the window xmin $= -2$, xmax $= 2$, xscl $= 1$, ymin $= -2$, ymax $= 2$, and Yscl $= 1$.

19. What three points do all of the graphs have in common?
20. Describe how the graphs are related.
21. Using what you have learned in Exercises 19 and 20, sketch the graph of $f(x) = \sqrt[5]{x+2} - 1$.
22. ***Life Expectancy*** From 1940 through 1996 in the United States, the age to which a newborn can expect to live can be modeled by

$$f(t) = 1.78\sqrt{t - 0.3} + 62.7,$$

where t is the number of years since 1940. Graph the model. In what year was the life expectancy at birth 75.7 years?

NAME ______________________________ DATE ____________

Reteaching with Practice

For use with pages 431–436

GOAL Graph square root and cube root functions

VOCABULARY

The graphs of $y = \sqrt{x}$ and $y = \sqrt[3]{x}$ are examples of **radical functions.** To graph $y = a\sqrt{x-h} + k$ or $a\sqrt[3]{x-h} + k$, follow these steps.

Step 1: Sketch the graph of $y = a\sqrt{x}$ or $a\sqrt[3]{x}$.

Step 2: Shift graph h units horizontally and k units vertically.

EXAMPLE 1 Comparing Two Graphs

Describe how to obtain the graph of $y = \sqrt[3]{x-2} + 4$ from the graph of $y = \sqrt[3]{x}$.

SOLUTION

Note that $y = \sqrt[3]{x-2} + 4$ is in the form $y = a\sqrt[3]{x-h} + k$, where $a = 1$, $h = 2$, and $k = 4$. To obtain the graph of $y = \sqrt[3]{x-2} + 4$, shift the graph of $y = \sqrt[3]{x}$ right 2 units and up 4 units.

Exercises for Example 1

Describe how to obtain the graph of *g* from the graph of *f*.

1. $g(x) = \sqrt{x} + 3, f(x) = \sqrt{x}$
2. $g(x) = \sqrt[3]{x+5}, f(x) = \sqrt[3]{x}$
3. $g(x) = \sqrt{x-1} - 4, f(x) = \sqrt{x}$
4. $g(x) = \sqrt[3]{x+1} + 6, f(x) = \sqrt[3]{x}$

EXAMPLE 2 Graphing a Square Root Function

Graph $y = -2\sqrt{x+3} + 5$.

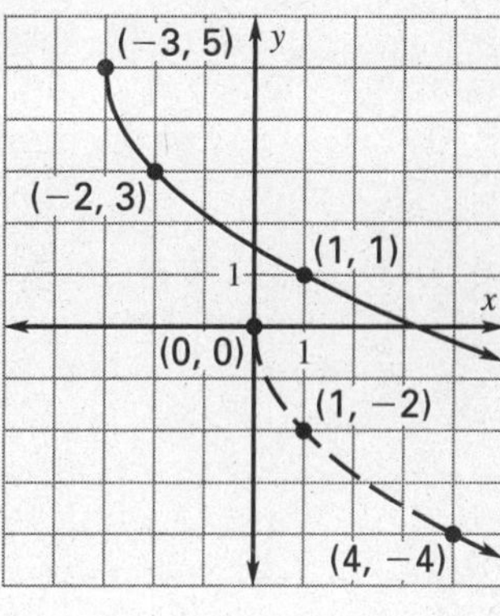

SOLUTION

Begin by sketching the graph of $y = -2\sqrt{x}$ (shown as a dashed curve). Notice that the graph begins at the origin and passes through the points $(1, -2)$ and $(4, -4)$.

Note that for $y = -2\sqrt{x+3} + 5$, $h = -3$ and $k = 5$. So, shift the graph left 3 units and up 5 units. Notice that the graph begins at $(-3, 5)$ and passes through the points $(1-3, -2+5) = (-2, 3)$ and $(4-3, -4+5) = (1, 1)$.

Exercises for Example 2

Graph the square root function.

5. $y = 4\sqrt{x}$
6. $y = 4\sqrt{x+3}$
7. $y = 4\sqrt{x} - 5$

NAME ______________________ DATE ____________

Reteaching with Practice

For use with pages 431–436

EXAMPLE 3 Graphing a Cube Root Function

Graph $y = -\sqrt[3]{x - 1} + 4$.

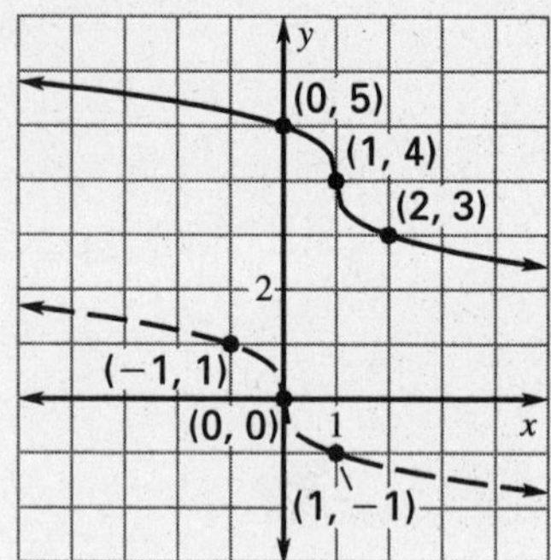

SOLUTION

Begin by sketching the graph of $y = -\sqrt[3]{x}$ (shown as a dashed curve). Notice that it passes through the origin and the points $(1, -1)$ and $(-1, 1)$. Note that for $y = -\sqrt[3]{x - 1} + 4$, $h = 1$ and $k = 4$. So, shift the graph right 1 unit and up 4 units. Notice that the graph passes through the points $(1, 4)$, $(2, 3)$, and $(0, 5)$.

Exercises for Example 3

Graph the cubic function.

8. $y = 3\sqrt[3]{x}$ **9.** $y = 3\sqrt[3]{x - 2}$ **10.** $y = 3\sqrt[3]{x} + 2$

EXAMPLE 4 Finding Domain and Range

State the domain and range of the function in (a) Example 2 and (b) Example 3.

SOLUTION

a. From the graph of $y = -2\sqrt{x + 3} + 5$ in Example 2, you can see that the graph begins at $x = -3$ and continues to the right. Therefore, the domain of the function is $x \geq -3$. The maximum value of y in the graph is $y = 5$. Therefore, the range of the function is $y \leq 5$.

b. From the graph of $y = -\sqrt[3]{x - 1} + 4$ in Example 3, you can see that the domain and range of the function are both all real numbers.

Exercises for Example 4

Using the graphs from Exercises 11–16, state the domain and range of the function.

11. $y = 4\sqrt{x}$ **12.** $y = 4\sqrt{x + 3}$ **13.** $y = 4\sqrt{x} - 5$

14. $y = 3\sqrt[3]{x}$ **15.** $y = 3\sqrt[3]{x - 2}$ **16.** $y = 3\sqrt[3]{x} + 2$

NAME ______________________________ DATE ________

Quick Catch-Up for Absent Students

For use with pages 431–436

The items checked below were covered in class on (date missed) __________

Lesson 7.5: Graphing Square Root and Cube Root Functions

____ **Goal 1:** Graph square root and cube root functions. (p. 431, 432)

Material Covered:

____ Activity: Investigating Graphs of Radical Functions

____ Example 1: Comparing Two Graphs

____ Student Help: Skills Review

____ Example 2: Graphing a Square Root Function

____ Example 3: Graphing a Cube Root Function

____ Example 4: Finding Domain and Range

Vocabulary:

radical function, p. 431

____ **Goal 2:** Use square root and cube root functions to find real-life quantities. (p. 433)

Material Covered:

____ Example 5: Modeling with a Square Root Function

____ Example 6: Modeling with a Cube Root Function

____ Other (specify) ______________________________

____

Homework and Additional Learning Support

____ Textbook (specify) pp. 434–436

____

____ *Reteaching with Practice* worksheet (specify exercises) __________

____ *Personal Student Tutor* for Lesson 7.5

NAME ______________________ DATE __________

Interdisciplinary Application

For use with pages 431–436

Body Mass Index

HEALTH The Body Mass Index (BMI) is used by health care physicians to give a quick prediction of body fat. The BMI compares a ratio relating a patient's weight and height. BMI is a better indicator than weight alone. BMI is used to alert patients who may be at risk of some health problems. Patients with a high BMI may have a higher chance of developing hypertension, adult diabetes, or cardiovascular disease.

BMI is used because it is a simple calculation and it applies to both men and women over the age of 18. However, children who are growing, pregnant women, and people with large amounts of muscle, such as competitive athletes, should not use BMI to estimate their body fat.

Most professionals consider BMI readings between 19.5 and 24.9 to be optimum. Readings between 18 and 19.5 may indicate a form of mild starvation. Readings over 30 may indicate high body fat content. While these may be recommended scores, the risks of higher body fat may be offset by a strong cardiovascular system. Your doctor can give the best advice to your specific body fat content. The BMI test should be used as an indicator.

In Exercises 1–5, use the following information.

A person's Body Mass Index M can be modeled by

$$h = 40\sqrt{\frac{0.45w}{M}}$$

where h is height in inches and w is weight in pounds.

1. Graph the model for a person who weighs 120 pounds.
2. Find the Body Mass Index of a patient who is 6 feet tall and weighs 170 pounds.
3. A person is 5 feet 5 inches tall and has a Body Mass Index reading of 25. Find the person's weight.
4. A physician suggests a person should have a Body Mass Index reading between 23 and 27. Find the two weights a person should try to stay between if he or she is 5 feet 9 inches tall.
5. Find the Body Mass Index readings of a person who is 6 feet 2 inches tall and weighed 210 pounds but now weighs 175 pounds.

LESSON 7.5

NAME ______________________ DATE ________

Challenge: Skills and Applications

For use with pages 431–436

In Exercises 1–2, graph each function. State the domain and the range of each.

1. $y = x^{3/2}$

2. $y = x^{2/3}$

3. On one set of axes, graph $y = \sqrt[3]{x}$ and $y = \sqrt{x}$ for $x \geq 0$.

a. At what point do the two graphs intersect?

b. For which graph are the y-coordinates of the points on the graph larger for $0 < x < 1$?

c. For which graph are the y-coordinates of the points on the graph larger for $x > 1$?

d. Graph the function $y = \sqrt[4]{x}$ on the same axes.

e. For any fixed x, what can you say about the values of $\sqrt[n]{x}$ as $n \to \infty$?

In Exercises 4–7, use a graphing calculator to find the intersection points of each pair of graphs.

4. $y = \sqrt[3]{x} + 1;\ y = \sqrt{x + 1}$

5. $y = \sqrt{x};\ y = \sqrt[3]{x} + 4$

6. $y = \sqrt[3]{x + 1};\ y = \sqrt{x - 1} - 2$

7. $y = \sqrt[3]{x + 8};\ y = 2\sqrt{1 - x}$

8. An "addition" operation can be set up using the points on the graph of $y = \sqrt[3]{x}$ as follows: To add points P and Q on the graph, let R be the 3rd point where the line PQ intersects the graph (if the line is tangent at Q, choose $R = Q$). Then $P + Q$ is defined to be $-R$, that is, the point whose coordinates are negatives of those of R.

a. Explain why the point (0, 0) is the identity element with respect to this addition.

b. What is the additive inverse of a point (a, b) on the graph? Explain.

c. Add the points (8, 2) and (27, 3). To do this, first find the equation of the line through these points. Then substitute for x in this equation using $y = \sqrt[3]{x}$; i.e. $x = y^3$. The cubic equation you get will have 3 zeros, two of which you already know.

LESSON
7.6

TEACHER'S NAME ____________________ CLASS ______ ROOM ______ DATE ______

Lesson Plan

2-day lesson (See *Pacing the Chapter,* TE pages 398C–398D) **For use with pages 437–444**

1. Solve equations that contain radicals or rational exponents.
2. Use radical equations to solve real-life problems.

State/Local Objectives ____

____

✓ Check the items you wish to use for this lesson.

STARTING OPTIONS
____ Homework Check: TE page 434; Answer Transparencies
____ Warm-Up or Daily Homework Quiz: TE pages 437 and 436, CRB page 78, or Transparencies

TEACHING OPTIONS
____ Motivating the Lesson: TE page 438
____ Lesson Opener (Graphing Calculator): CRB page 79 or Transparencies
____ Graphing Calculator Activity with Keystrokes: CRB pages 80–81
____ Examples: Day 1: 1–4, SE pages 437–438; Day 2: 5–6, SE pages 439–440
____ Extra Examples: Day 1: TE page 438 or Transp.; Day 2: TE pages 439–440 or Transp.; Internet
____ Closure Question: TE page 440
____ Guided Practice: SE page 441 Day 1: Exs. 2–9; Day 2: Exs. 1, 10–16

APPLY/HOMEWORK
Homework Assignment
____ Basic Day 1: 17–22, 24–42 even, 63–64; Day 2: 48–52 even, 56–60 even, 70, 71–73, 81–89 odd
____ Average Day 1: 17–22, 24–46 even, 56–60 even, 63–65; Day 2: 48–54 even, 61, 62, 69–73, 81–89
____ Advanced Day 1: 17–22, 24–56 even, 63–65; Day 2: 51–61 odd, 66–79, 81–89

Reteaching the Lesson
____ Practice Masters: CRB pages 82–84 (Level A, Level B, Level C)
____ Reteaching with Practice: CRB pages 85–86 or Practice Workbook with Examples
____ Personal Student Tutor

Extending the Lesson
____ Cooperative Learning Activity: CRB page 88
____ Applications (Real-Life): CRB page 89
____ Math & History: SE page 444; CRB page 90; Internet
____ Challenge: SE page 443; CRB page 91 or Internet

ASSESSMENT OPTIONS
____ Checkpoint Exercises: Day 1: TE page 438 or Transp.; Day 2: TE pages 439–440 or Transp.
____ Daily Homework Quiz (7.6): TE page 443, CRB page 94, or Transparencies
____ Standardized Test Practice: SE page 443; TE page 443; STP Workbook; Transparencies

Notes ____

____

____

LESSON 7.6

TEACHER'S NAME ______________________ CLASS ______ ROOM ______ DATE ______

Lesson Plan for Block Scheduling

1-day lesson (See *Pacing the Chapter,* TE pages 398C–398D) **For use with pages 437–444**

GOALS

1. **Solve equations that contain radicals or rational exponents.**
2. **Use radical equations to solve real-life problems.**

State/Local Objectives ______________________

____

____

CHAPTER PACING GUIDE	
Day	**Lesson**
1	Assess Ch. 6; 7.1(all)
2	7.2 (all)
3	7.3 (all); 7.4(begin)
4	7.4 (end); 7.5(all)
5	**7.6 (all)**
6	7.7 (all)
7	Review/Assess Ch. 7

✓ Check the items you wish to use for this lesson.

STARTING OPTIONS

____ Homework Check: TE page 434; Answer Transparencies

____ Warm-Up or Daily Homework Quiz: TE pages 437 and 436, CRB page 78, or Transparencies

TEACHING OPTIONS

____ Motivating the Lesson: TE page 438

____ Lesson Opener (Graphing Calculator): CRB page 79 or Transparencies

____ Graphing Calculator Activity with Keystrokes: CRB pages 80–81

____ Examples: 1–6: SE pages 437–440

____ Extra Examples: TE pages 438–440 or Transparencies; Internet

____ Closure Question: TE page 440

____ Guided Practice Exercises: SE page 441

APPLY/HOMEWORK

Homework Assignment

____ Block Schedule: 17–22, 24–60 even, 63–65, 69–73, 81–89

Reteaching the Lesson

____ Practice Masters: CRB pages 82–84 (Level A, Level B, Level C)

____ Reteaching with Practice: CRB pages 85–86 or Practice Workbook with Examples

____ Personal Student Tutor

Extending the Lesson

____ Cooperative Learning Activity: CRB page 88

____ Applications (Real-Life): CRB page 89

____ Math & History: SE page 444; CRB page 90; Internet

____ Challenge: SE page 443; CRB page 91 or Internet

ASSESSMENT OPTIONS

____ Checkpoint Exercises: TE pages 438–440 or Transparencies

____ Daily Homework Quiz (7.6): TE page 443, CRB page 94, or Transparencies

____ Standardized Test Practice: SE page 443; TE page 443; STP Workbook; Transparencies

Notes ______________________

____

____

LESSON **7.6**

NAME ______________________ DATE __________

Available as a transparency

WARM-UP EXERCISES

For use before Lesson 7.6, pages 437–444

1. Multiply $(x - 3)^2$.

2. Factor $x^2 + 8x + 16$.

Evaluate when $x = 4$.

3. $\sqrt{10x - 4}$

4. $\sqrt[3]{3x - 4}$

5. $\sqrt[3]{6x + 192} - \sqrt[3]{2x}$

Lesson 7.6

DAILY HOMEWORK QUIZ

For use after Lesson 7.5, pages 431–436

Describe how to obtain the graph of *g* from the graph of *f*.

1. $g(x) = 3\sqrt{x + 2} - 1$

$f(x) = 3\sqrt{x}$

Graph the function. Then state the domain and range.

2. $y = 3x^{1/2}$

3. $y = \sqrt[3]{x} + 4$

NAME ______________________ DATE ________

Available as a transparency

Graphing Calculator Lesson Opener

For use with pages 437–444

In Lesson 7.6, you will learn algebraic methods for solving equations that contain radicals or rational exponents. You can obtain approximate solutions to these equations using a graphing calculator.

For example, to solve $\sqrt[3]{x-2} = 2 - 1.5x$, graph $y_1 = \sqrt[3]{x-2}$ and $y_2 = 2 - 1.5x$. Use the *Intersect* feature. The solution is $x \approx 1.75$.

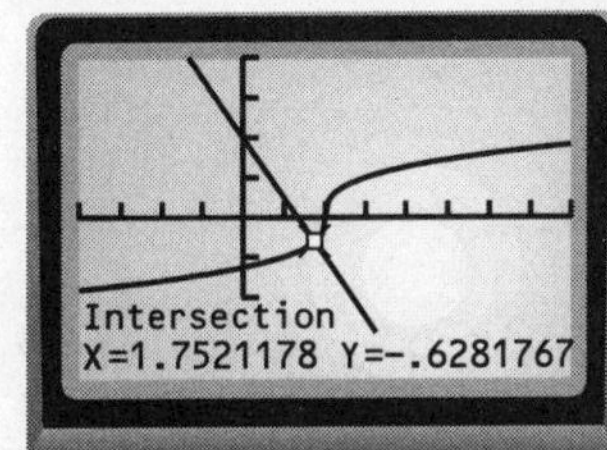

Keep in mind that sometimes an equation will have more than one solution or will not have any solutions.

Use a graphing calculator to solve the equation. If necessary, round to the nearest hundredth.

1. $\sqrt{3-x} = 2x + 1$

2. $(2x+5)^{1/2} = 0.5x + 2$

3. $\sqrt[3]{x+3} = 3 - 5x$

4. $5 - x = \sqrt[3]{4x-3}$

5. $x^{1/3} = 0.25x + 0.5$

6. $1 - 0.5x = \sqrt[3]{2-x}$

7. $\sqrt{5-x} = \sqrt{x+2}$

8. $\sqrt{2x+5} + \sqrt{20-4x} = 6$

LESSON 7.6

NAME ______________________ DATE __________

Graphing Calculator Activity

For use with pages 437–444

GOAL **To solve equations that contain radical expressions**

Activity

1. Graph $y = (x^{1/3})^3$ to see that it is the same as $y = x$.
2. Graph $y = (x^{3/2})^{2/3}$ to see that it is the same as $y = x$.
3. To solve $x^{1/3} = 2$, use a graphing calculator to find the intersection of $y = x^{1/3}$ and $y = 2$.

Lesson 7.6

Exercises

1. Solve $x^{1/5} = 2$ algebraically by raising both sides to the fifth power.

2. Verify your solution to Exercise 1 by using a graphing calculator to find the intersection of $y = x^{1/5}$ and $y = 2$.

3. Solve $2x^{1/4} = 6$ algebraically as follows. Isolate x by dividing both sides by 2. Then raise both sides to the fourth power.

4. Verify your solution to Exercise 3 by using a graphing calculator to find the intersection of $y = 2x^{1/4}$ and $y = 6$.

5. Solve $\left(\frac{2}{5}\right)x^{1/2} = 10$ algebraically as follows. Isolate x by multiplying both sides by $\frac{5}{2}$. Then raise both sides to the second power.

6. Verify your solution to Exercise 5 by using a graphing calculator to find the intersection of $y = \left(\frac{2}{5}\right)x^{1/2}$ and $y = 10$.

NAME ______________________________ DATE ____________

Graphing Calculator Activity

For use with pages 437–444

TI-82

Step 1

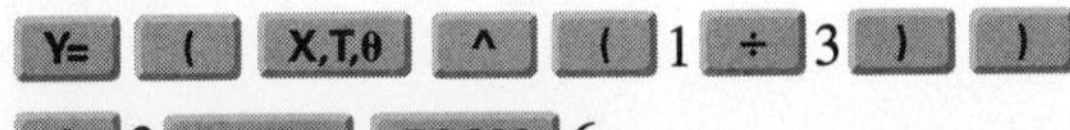

^ 3 ENTER ZOOM 6

Step 2

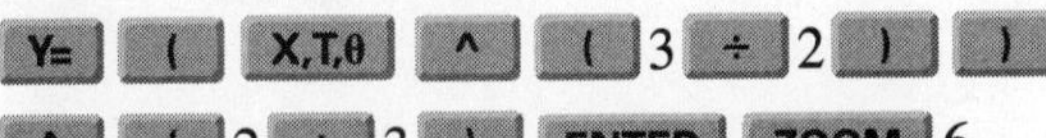

^ (2 ÷ 3) ENTER ZOOM 6

Step 3

ENTER ZOOM 6

Find the point of intersection at $x = 8$.

2nd [CALC]5 ENTER ENTER

Use the cursor keys to select the guess near $x = 8$.

ENTER

TI-83

Step 1

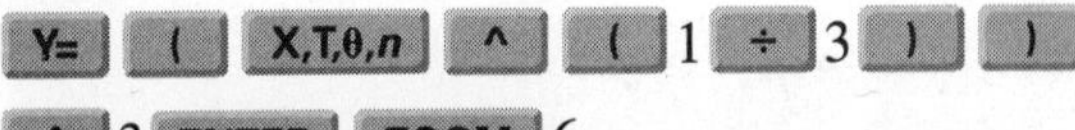

^ 3 ENTER ZOOM 6

Step 2

Y= (X,T,θ,n ^ (3 ÷ 2))

^ (2 ÷ 3) ENTER ZOOM 6

Step 3

Y= X,T,θ,n ENTER 2 ENTER ZOOM 6

Find the point of intersection at $x = 8$.

2nd [CALC]5 ENTER ENTER 8 ENTER

SHARP EL-9600c

Step 1

Y= (X/θ/T/n a^b 1 ÷ 3 ▶)

^ 3 ENTER ZOOM [A]5

Step 2

Y= (X/θ/T/n a^b 3 ÷ 2 ▶) a^b

2 ÷ 3 ▶ ENTER ZOOM [A]5

Step 3

Y= X/θ/T/n a^b 1 ÷ 3 ▶ ENTER 2

ENTER ZOOM [A]5

Find the point of intersection at $x = 8$.

2nd [CALC]2

CASIO CFX-9850GA PLUS

From the main menu, choose GRAPH.

Step 1

(X,θ,T ^ (1 ÷ 3))

^ 3 EXE SHIFT F3 F3 EXIT F6

Step 2

(X,θ,T ^ (3 ÷ 2))

^ (2 ÷ 3) EXE SHIFT F3 F3

EXIT F6

Step 3

X,θ,T ^ (1 ÷ 3) EXE EXE

SHIFT F3 F3 EXIT F6

Find the point of intersection at $x = 8$.

SHIFT F5 F5

LESSON **7.6**

NAME ______________________ DATE __________

Practice A

For use with pages 437–444

Check whether the given x-value is a solution of the equation.

1. $\sqrt{x} - 5 = 8;\ x = 169$
2. $\sqrt{2x - 1} + 2 = 5;\ x = 5$
3. $\sqrt{x} + 4 = 10;\ x = 25$
4. $\sqrt{1 - x} + 3 = 5;\ x = -3$
5. $2\sqrt{x + 5} = 12;\ x = 20$
6. $\sqrt[3]{x + 1} - 3 = -2;\ x = 0$

Solve the equation. Check for extraneous solutions.

7. $x^{1/4} = 2$
8. $x^{2/3} = 16$
9. $x^{1/2} = 8$
10. $x^{1/3} - 2 = 0$
11. $x^{3/2} + 4 = 12$
12. $4x^{2/3} = 100$

Solve the equation. Check for extraneous solutions.

13. $\sqrt{x} = \frac{1}{4}$
14. $\sqrt[3]{x} = -2$
15. $\sqrt[4]{x^3} = 27$
16. $\sqrt[5]{x} + 3 = 4$
17. $2\sqrt[3]{x} = 6$
18. $5\sqrt[3]{x} = -15$

Solve the equation. Check for extraneous solutions.

19. $\sqrt{x + 3} = \sqrt{6}$
20. $\sqrt{2x + 1} = \sqrt{x}$
21. $\sqrt{3x - 2} = \sqrt{2}$
22. $\sqrt[3]{x - 5} = \sqrt[3]{5}$
23. $\sqrt[4]{3x - 1} = \sqrt[4]{2x + 2}$
24. $\sqrt[3]{5x - 6} = \sqrt[3]{4}$

Pendulums **In Exercises 25–27, use the following information.**

The period of a pendulum is the time T (in seconds) it takes for a pendulum of length L (in feet) to go through one cycle. The period is given by

$$T = 2\pi\sqrt{\frac{L}{32}}.$$

Given the period of a pendulum, find its length. Round your answers to two decimal places.

25. $T = 1$ second
26. $T = 0.5$ second
27. $T = 2$ seconds

Velocity of a Free-Falling Object **In Exercises 28–30, use the following information.**

The velocity of a free-falling object is given by $V = \sqrt{2gh}$, where V is velocity (in feet per second), g is acceleration due to gravity (in feet per second) and h is the distance (in feet) the object has fallen. On Earth $g = 32$ ft/s^2. How far did an object fall if it hits the ground with the given velocity?

28. 80 ft/s
29. 48 ft/s
30. 120 ft/s

LESSON **7.6**

NAME ______________________ DATE ____________

Practice B

For use with pages 437–444

Solve the equation. Check for extraneous solutions.

1. $x^{4/3} - 5 = 11$
2. $2x^{3/4} + 7 = 23$
3. $(2x)^{3/4} = 8$
4. $(x - 1)^{2/3} = 4$
5. $2(x + 1)^{3/2} = 54$
6. $2x^{5/3} = -64$
7. $(2x + 3)^{1/3} - 5 = -2$
8. $(2x - 1)^{1/5} + 2 = 3$
9. $-(3x + 4)^{1/2} + 3 = 0$

Solve the equation. Check for extraneous solutions.

10. $\sqrt[4]{3x} + 5 = 6$
11. $3\sqrt{x + 6} + 5 = 14$
12. $\sqrt{5x - 1} + 8 = 2$
13. $\sqrt[3]{2x + 1} + 2 = 4$
14. $-\sqrt[3]{5x + 4} + 1 = -3$
15. $\sqrt[3]{3x + 1} + 5 = 3$
16. $\sqrt[5]{3 - x} + 4 = 3$
17. $2\sqrt[3]{1 - 3x} + 4 = 6$
18. $5 - \sqrt{2x + 1} = 3$

Solve the equation. Check for extraneous solutions.

19. $\sqrt[3]{2x + 1} = \sqrt[3]{8}$
20. $\sqrt{3x + 1} = \sqrt{x - 5}$
21. $\sqrt[4]{2x + 1} = \sqrt[4]{x + 6}$
22. $\sqrt{x + 2} = x + 2$
23. $\sqrt{2x - 3} = x - 3$
24. $\sqrt{12x + 13} = 2x + 1$
25. $\sqrt{3x + 13} = x + 5$
26. $\sqrt{2x} = x - 4$
27. $2\sqrt{x + 4} - 1 = x$

Use the *Intersect* feature on a graphing calculator to solve the equation.

28. $\frac{2}{3}x^{1/2} = 1$
29. $6(x + 3)^{3/5} = 18$
30. $(2x + 5)^{1/3} = -2$
31. $\sqrt{1.3x + 11} = 4$
32. $\sqrt[3]{43 - 5x} = 2.1$
33. $(2x + 3)^{2/3} = 3$

Velocity of a Free-Falling Object **In Exercises 34–36, use the following information.**

The velocity of a free-falling object is given by $V = \sqrt{2gh}$ where h is the distance (in feet) the object has fallen and g is acceleration due to gravity (in feet per second squared). The value of g depends on your altitude. If an object hits the ground with a velocity of 25 feet per second, from what height was it dropped in each of the following situations?

34. You are standing on the earth, so $g = 32\ \text{ft/s}^2$.
35. You are on the space shuttle, so $g = 29\ \text{ft/s}^2$.
36. You are on the moon, so $g = 0.009\ \text{ft/s}^2$.

NAME ______________________ DATE ________

Practice C

For use with pages 437–444

Solve the equation. Check for extraneous solutions.

1. $3(x-1)^{2/3}+4=52$
2. $2(x+4)^{1/3}+7=-9$
3. $-(2x+3)^{2/3}+5=1$
4. $\frac{1}{2}(3x-1)^{3/4}-3=1$
5. $\frac{1}{3}(2x+3)^{3/2}+2=-7$
6. $\frac{1}{3}(2x+3)^{3/2}-2=7$

Solve the equation. Check for extraneous solutions.

7. $3\sqrt{\frac{1}{2}x-5}+1=7$
8. $4-\sqrt{3x+1}=5$
9. $\frac{1}{5}\sqrt[3]{2x-\frac{1}{2}}+3=6$
10. $\sqrt{x^2+3}-5=4$
11. $2\sqrt{x^2-1}+4=10$
12. $3\sqrt[3]{1-x^2}+1=-8$

Solve the equation. Check for extraneous solutions.

13. $\sqrt[5]{3x+7}=\sqrt[5]{2x+1}$
14. $\sqrt{\frac{2}{3}+x}=-\sqrt{2x+\frac{1}{3}}$
15. $\sqrt{x-7}=x-7$
16. $\sqrt{3x^2-12x+10}=2x-5$
17. $\sqrt[4]{2x^2-1}=x$
18. $\sqrt[3]{9x+19}=x+1$
19. $\sqrt[3]{2x^2+14}=x-1$
20. $\sqrt[5]{4x^3+x^2-4}=x$
21. $-\sqrt{x-3}=x-5$

Solve the equation. Check for extraneous solutions.

22. $\sqrt{x+3}=4-\sqrt{x}$
23. $\sqrt{x-5}=2+\sqrt{x}$
24. $\sqrt{x-5}=2-\sqrt{x}$
25. $\sqrt{5x+1}=3-\sqrt{5x}$
26. $\sqrt{2x+1}=1+\sqrt{2x}$
27. $\sqrt{2x+3}=1+\sqrt{x+1}$

28. ***Geometry*** The lateral surface area of a cone is given by $S=\pi r\sqrt{r^2+h^2}$. The surface area of the base of the cone is given by $B=\pi r^2$. The total surface area of a cone of radius 3 inches is 24π square inches. What is the height of the cone?

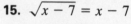

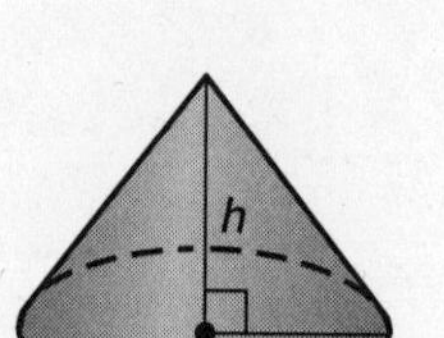

29. ***Geometry*** A container is to be made in the shape of a cylinder with a conical top. The lateral surface areas of the cylinder and cone are $S_1=2\pi rh$ and $S_2=2\pi r\sqrt{r^2+h^2}$. The surface area of the base of the container is $B=\pi r^2$. The height of the cylinder and cone are equal. The radius of the container is 5 inches and its total surface area is 275π square inches. Find the total height of the container.

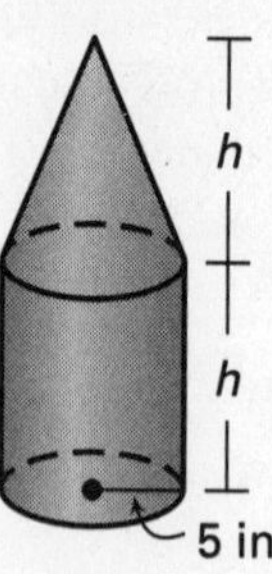

NAME ______________________________ DATE ____________

Reteaching with Practice

For use with pages 437–444

GOAL Solve equations that contain radicals or rational exponents

VOCABULARY

The **powers property of equality** states that if $a = b$, then $a^n = b^n$. In other words, you can raise each side of an equation to the same power. An **extraneous solution** is a trial solution that does not satisfy the original equation.

EXAMPLE 1 Solving a Simple Radical Equation

Solve $\sqrt{x} + 5 = 9$.

SOLUTION

$\sqrt{x} + 5 = 9$	Write original equation.
$\sqrt{x} = 4$	Isolate the radical by subtracting 5 from each side.
$(\sqrt{x})^2 = 4^2$	Square each side.
$x = 16$	Simplify.

The solution is 16. You can check this substituting 16 for x in the original equation to get $\sqrt{16} + 5 = 4 + 5 = 9$.

Exercises for Example 1

Solve the equation. Check your solution.

1. $\sqrt[3]{x} + 2 = 0$

2. $-\sqrt{x} - 5 = -6$

3. $\sqrt[4]{x} = 3$

EXAMPLE 2 Solving an Equation with Rational Exponents

Solve $3x^{3/4} = 192$.

SOLUTION

$3x^{3/4} = 192$	Write original equation.
$x^{3/4} = 64$	Isolate the power by dividing each side by 3.
$(x^{3/4})^{4/3} = 64^{4/3}$	Raise each side to $\frac{4}{3}$ power, the reciprocal of $\frac{3}{4}$.
$x = (64^{1/3})^4$	Apply properties of roots.
$x = 4^4 = 256$	Simplify.

The solution is 256. You can check this by substituting 256 for x in the original equation to get $3(256)^{3/4} = 3[(256)^{1/4}]^3 = 3(4)^3 = 3(64) = 192$.

NAME ______________________ DATE __________

Reteaching with Practice

For use with pages 437–444

Exercises for Example 2

Solve the equation. Check your solution.

4. $2x^{1/2} = 18$

5. $5x^{3/2} = 40$

6. $x^{3/4} = \frac{1}{8}$

EXAMPLE 3 Solving an Equation with One Radical

Solve $\sqrt[3]{8x + 3} - 5 = -2$.

SOLUTION

$\sqrt[3]{8x + 3} - 5 = -2$	Write original equation.
$\sqrt[3]{8x + 3} = 3$	Isolate the radical, by adding 5 to each side.
$\left(\sqrt[3]{8x + 3}\right)^3 = 3^3$	Cube each side.
$8x + 3 = 27$	Simplify.
$8x = 24$	Subtract 3 from each side.
$x = 3$	Divide each side by 8.

The solution is 3. Check this in the original equation.

Exercises for Example 3

Solve the equation. Check your solution.

7. $\sqrt{4 + 3x} = 10$

8. $\sqrt{2x + 1} = 7$

9. $\sqrt[3]{4x - 1} = 3$

EXAMPLE 4 Solving an Equation with Two Radicals

Solve $\sqrt[3]{2x + 4} = 2\sqrt[3]{3 - x}$.

SOLUTION

$\sqrt[3]{2x + 4} = 2\sqrt[3]{3 - x}$	Write original equation.
$\left(\sqrt[3]{2x + 4}\right)^3 = \left(2\sqrt[3]{3 - x}\right)^3$	Cube each side.
$2x + 4 = 8(3 - x)$	Simplify.
$2x + 4 = 24 - 8x$	Distributive property
$10x + 4 = 24$	Add $8x$ to each side.
$10x = 20$	Subtract 4 from each side.
$x = 2$	Divide each side by 10.

The solution is 2. Check this in the original equation.

Exercises for Example 4

Solve the equation. Check your solution.

10. $\sqrt{7x - 8} = \sqrt{5x}$

11. $\sqrt{3x + 5} = \sqrt{x + 15}$

12. $\sqrt[3]{x + 14} = 2\sqrt[3]{x}$

NAME ______________________________ DATE __________

Quick Catch-Up for Absent Students

For use with pages 437–444

The items checked below were covered in class on (date missed) __________

Lesson 7.6: Solving Radical Equations

____ **Goal 1:** Solve equations that involve radicals or rational exponents. (p. 437–439)

Material Covered:

____ Example 1: Solving a Simple Radical Equation

____ Student Help: Study Tip

____ Example 2: Solving an Equation with Rational Exponents

____ Example 3: Solving an Equation with One Radical

____ Example 4: Solving an Equation with Two Radicals

____ Student Help: Look Back

____ Example 5: An Equation with an Extraneous Solution

Vocabulary:

extraneous solution, p. 439

____ **Goal 2:** Use radical equations to solve real-life problems. (p. 440)

Material Covered:

____ Example 6: Using a Radical Model

____ Other (specify) ______________________________

____

Homework and Additional Learning Support

____ Textbook (specify) pp. 441–444

____

____ Internet: Extra Examples at www.mcdougallittell.com

____ *Reteaching with Practice* worksheet (specify exercises) __________

____ *Personal Student Tutor* for Lesson 7.6

LESSON 7.6

NAME ______________________________ DATE ________

Cooperative Learning Activity

For use with pages 437–444

Lesson 7.6

GOAL **To solve a radical equation and find the time it takes a falling object to hit the ground**

Materials: paper, calculator

Background

The number of seconds it takes for a bowling ball to hit the ground when the bowling ball is dropped from a height of 80 meters can be modeled by

$$t = \sqrt{\frac{80 - s}{4.9}}$$

where t is the time in seconds and s is the height (in meters) of the bowling ball.

Instructions

1. Use the model given to complete the following table.

Time, t	0	1	2	3	4	5
Distance, s						

2. Substitute 0 for s in the model above and solve for t.

Analyzing the Results

1. When the height s of the bowling ball is 0 meters, the bowling ball has hit the ground. According to the table, after how many seconds does the bowling ball hit the ground? What answer did you get in Step 2?

2. In Step 1, you constructed a table and were able to use a numerical approach to estimate the time it took for the bowling ball to hit the ground. In Step 2, you used an algebraic approach to find this time. Which approach did you prefer? Why?

3. Suppose the bowling ball is dropped again, this time from a height of 160 meters. Will it take the bowling ball twice as long to reach the ground? Use the model

$$t = \sqrt{\frac{160 - s}{4.9}}$$

to answer the question.

NAME ______________________ DATE __________

Real-Life Application: When Will I Ever Use This?

For use with pages 437–444

Airspeed

Airplane speeds are measured in three different ways: indicated speed, true speed, and ground speed. The indicated airspeed is the airspeed given by an instrument called an airspeed indicator. A plane's indicated airspeed is different from its true airspeed because the indicator is affected by temperature changes and different altitudes of air pressure. The true airspeed is the speed of the airplane relative to the wind. Ground speed is the speed of the airplane relative to the ground. For example, a plane flying at a true airspeed of 150 knots into a headwind of 25 knots will have a ground speed of 125 knots.

The exercises below refer to static and dynamic pressure. Static pressure is used when a body is in motion or at rest at a constant speed and direction. Dynamic pressure is used when a body in motion changes speed or direction or both. A gauge compares these pressures, giving pilots an indicated airspeed.

In Exercises 1 and 2, use the following information.

The indicated airspeed S (in knots) of an airplane is given by an airspeed indicator that measures the difference p (in inches of mercury) between the static and dynamic pressures. The relationship between S and p can be modeled by

$$S = 136.4\sqrt{p} + 4.5.$$

1. Find the differential pressure when the indicated airspeed is 157 knots.
2. Find the change in the differential pressure of an airplane that was traveling at 218 knots and slowed down to 195 knots.

In Exercises 3 and 4, use the following information.

The true airspeed T (in knots) of an airplane can be modeled by

$T = \left(1 + \dfrac{A}{50{,}000}\right)S$, where A is the altitude (in feet) and S is the indicated airspeed (in knots).

3. Write the equation for true airspeed T in terms of altitude and differential pressure p.
4. A plane is flying with a true airspeed of 279.7 knots at an altitude of 20,000 feet. Estimate the differential pressure.

NAME ______________________ DATE ________

Math and History Application

For use with page 444

HISTORY Oceanographers classify the tsunami as a shallow-water wave. This may seem strange, since a tsunami may travel over the deepest parts of the Pacific, but the classification really means that the wavelength of the wave is long compared to the depth of the water. Shallow-water waves have wavelengths at least 20 times the depth of the ocean over which they are traveling. For example, a tsunami traveling over a region where the ocean depth is 2 km might have a wavelength of 180 km, and since 180 is more than $20 \cdot 2$ the wave behaves as a shallow-water wave.

If you were on a ship when the wave passed you might not even notice it, since over deep water the height of the wave might be as small as 1 foot. But, according to the formula $s = 356\sqrt{d}$ given on page 444, the wave will slow down as it approaches shore and the depth d gets smaller. To maintain its energy, the wave compensates for the slower speed by getting higher, so that by the time it hits shore the wave may have a height of 10 or even 100 feet.

Because these waves can be so devastating when they reach shore, there is now an active program that monitors all of the seismic activity that could produce a tsunami, and warnings are published on the Internet whenever a potentially dangerous event occurs.

MATH These problems explore some of the relationships between speed, wavelength, and period for ocean waves. Remember that in the formula on page 444 the speed s is in km/hour and the depth d is in km. The *period* of a wave is the time between crests if you stay in one spot as the wave moves by you. If we let P stand for the period in hours and L stand for the wavelength in kilometers, then the speed s is given by the formula $s = \frac{L}{P}$.

1. How fast in km/hour will a shallow-water wave travel over an ocean that is 4 km deep?

2. Suppose that a tsunami is traveling at 400 km/hour as it passes a certain point in the Pacific Ocean. How deep is the ocean at that point?

3. If a wave with wavelength 100 km is traveling at 200 km/hour, what is its period in minutes?

4. For deep-water waves, the wavelength is less than the depth of the ocean, so the wave doesn't "feel" the bottom and the speed does not depend on the depth as it does for the tsunami. For deep-water waves, like the ordinary ocean swell that you feel on a fishing boat, the speed s is entirely determined by the wavelength L according to the formula $s = 1.25\sqrt{L}$. Here we measure s in meters per second and L in meters. Find the speed of a swell with a wavelength of 10 meters.

5. Use the formula in problem 4 to find the wavelength of a deep-water wave traveling at 12 meters per second.

NAME ______________________________ DATE __________

Challenge: Skills and Applications

For use with pages 437–444

In Exercises 1–12, solve the equation.

Example $\sqrt{x+1} + \sqrt{x-2} = 3$

Solution $\sqrt{x+1} + \sqrt{x-2} = 3$

$$\sqrt{x+1} = 3 - \sqrt{x-2}$$

$$x + 1 = 9 + 6\sqrt{(x-2)} + (x-2)$$

$$-6 = 6\sqrt{x-2}$$

$$-1 = \sqrt{x-2}$$

$$1 = x - 2$$

$$x = 3$$

Check: $\sqrt{3+1} + \sqrt{3-2} = \sqrt{4} + \sqrt{1} = 2 + 1 = 3$

1. $\sqrt{2x-1} = 5 - \sqrt{x-1}$

2. $\sqrt{3-x} + \sqrt{x+5} = 4$

3. $\sqrt{7-x} - \sqrt{x-6} = -1$

4. $\sqrt{5x+1} = \sqrt{x-2} + 3$

5. $\sqrt[3]{x^3-7} + 1 = x$

6. $x + 2 = \sqrt[3]{x^3+8}$

7. $\sqrt{x-5} = \sqrt{x} - \sqrt{2}$

8. $\sqrt{3} + \sqrt{x-2} = \sqrt{x+3}$

9. $\frac{1}{2}\sqrt{3x+6} = 1 + \sqrt{x-6}$

10. $\sqrt{x} + \sqrt{x-5} = \sqrt{2x+7}$

11. $\dfrac{2x-5}{\sqrt{x-1}} = \sqrt{x-1}$

12. $\dfrac{4x-5}{\sqrt{2x+5}} = \sqrt{2x} - 1$

13. $\sqrt[3]{x+4} = \sqrt{x}$ (*Hint:* Start by cubing both sides.)

14. Suppose you tried to solve the equation $x - 1 = 2$ by squaring both sides.

a. Write the quadratic equation you get, and solve for x.

b. Check your solutions in the original equation, $x - 1 = 2$.

c. Where in your solution was an extraneous root introduced? Explain how.

LESSON 7.7

TEACHER'S NAME ______________________ CLASS ______ ROOM ______ DATE ______

Lesson Plan

2-day lesson (See *Pacing the Chapter,* TE pages 398C–398D)

For use with pages 445–454

GOALS

1. **Use measures of central tendency and measures of dispersion to describe data sets.**
2. **Use box-and-whisker plots and histograms to represent data graphically.**

State/Local Objectives ____

____

✓ Check the items you wish to use for this lesson.

STARTING OPTIONS

____ Homework Check: TE page 441; Answer Transparencies
____ Warm-Up or Daily Homework Quiz: TE pages 445 and 443, CRB page 94, or Transparencies

TEACHING OPTIONS

____ Motivating the Lesson: TE page 446
____ Lesson Opener (Application): CRB page 95 or Transparencies
____ Graphing Calculator Activity with Keystrokes: CRB page 96
____ Examples: Day 1: 1–2, SE pages 445–446; Day 2: 3–6, SE pages 446–448
____ Extra Examples: Day 1: TE page 446 or Transp.; Day 2: TE pages 446–448 or Transp.; Internet
____ Technology Activity: SE pages 453–454
____ Closure Question: TE page 448
____ Guided Practice: SE page 449 Day 1: Exs. 1, 4–5; Day 2: Exs. 2–3, 6–9

APPLY/HOMEWORK

Homework Assignment

____ Basic Day 1: 10–30 even, 38; Day 2: 32–34, 39–41, 46–48, 51–65 odd; Quiz 3: 1–12
____ Average Day 1: 10–30 even, 31, 38; Day 2: 32–37, 39–44, 46–48, 51–65 odd; Quiz 3: 1–12
____ Advanced Day 1: 10–30 even, 31–34, 38; Day 2: 35–37, 39–41, 43–49, 51–65 odd; Quiz 3: 1–12

Reteaching the Lesson

____ Practice Masters: CRB pages 97–99 (Level A, Level B, Level C)
____ Reteaching with Practice: CRB pages 100–101 or Practice Workbook with Examples
____ Personal Student Tutor

Extending the Lesson

____ Applications (Interdisciplinary): CRB page 103
____ Challenge: SE page 451; CRB page 104 or Internet

ASSESSMENT OPTIONS

____ Checkpoint Exercises: Day 1: TE page 446 or Transp.; Day 2: TE pages 446–448 or Transp.
____ Daily Homework Quiz (7.7): TE page 452 or Transparencies
____ Standardized Test Practice: SE page 451; TE page 452; STP Workbook; Transparencies
____ Quiz (7.5–7.7): SE page 452

Notes ____

____

____

LESSON 7.7

TEACHER'S NAME ______________________ CLASS ______ ROOM ______ DATE ______

Lesson Plan for Block Scheduling

1-day lesson (See *Pacing the Chapter,* TE pages 398C–398D) **For use with pages 445–454**

GOALS

1. **Use measures of central tendency and measures of dispersion to describe data sets.**
2. **Use box-and-whisker plots and histograms to represent data graphically.**

State/Local Objectives ______________________________

____

____

CHAPTER PACING GUIDE	
Day	**Lesson**
1	Assess Ch. 6; 7.1(all)
2	7.2 (all)
3	7.3 (all); 7.4(begin)
4	7.4 (end); 7.5(all)
5	7.6 (all)
6	**7.7 (all)**
7	Review/Assess Ch. 7

✓ Check the items you wish to use for this lesson.

STARTING OPTIONS

____ Homework Check: TE page 441; Answer Transparencies
____ Warm-Up or Daily Homework Quiz: TE pages 445 and 443, CRB page 94, or Transparencies

TEACHING OPTIONS

____ Motivating the Lesson: TE page 446
____ Lesson Opener (Application): CRB page 95 or Transparencies
____ Graphing Calculator Activity with Keystrokes: CRB page 96
____ Examples: 1–6: SE pages 445–448
____ Extra Examples: TE pages 446–448 or Transparencies; Internet
____ Technology Activity: SE pages 453–454
____ Closure Question: TE page 448
____ Guided Practice Exercises: SE page 449

APPLY/HOMEWORK

Homework Assignment

____ Block Schedule: 10–30 even, 31, 32–37, 39–44, 46–48, 51–65 odd; Quiz 3: 1–12

Reteaching the Lesson

____ Practice Masters: CRB pages 97–99 (Level A, Level B, Level C)
____ Reteaching with Practice: CRB pages 100–101 or Practice Workbook with Examples
____ Personal Student Tutor

Extending the Lesson

____ Applications (Interdisciplinary): CRB page 103
____ Challenge: SE page 451; CRB page 104 or Internet

ASSESSMENT OPTIONS

____ Checkpoint Exercises: TE pages 446–448 or Transparencies
____ Daily Homework Quiz (7.7): TE page 452 or Transparencies
____ Standardized Test Practice: SE page 451; TE page 452; STP Workbook; Transparencies
____ Quiz (7.5–7.7): SE page 452

Notes ______________________________

____

____

LESSON 7.7

NAME ______________________________ DATE __________

Available as a transparency

Warm-Up Exercises

For use before Lesson 7.7, pages 445–454

Evaluate the expression. Express your answer to the nearest hundredth.

1. $\dfrac{6 + 9 + 12 + 15}{4}$

2. $(15 - 10.5)^2$

3. $\sqrt{72.25}$

4. $\sqrt{\dfrac{328}{12}}$

5. $\sqrt{\dfrac{(12 - 10.5)^2 + (15 - 10.5)^2}{2}}$

Lesson 7.7

Daily Homework Quiz

For use after Lesson 7.6, pages 437–444

Solve the equation. Check for extraneous solutions.

1. $x^{2/3} + 4 = 20$

2. $(3x)^{1/2} - 7 = 2$

3. $\sqrt{x - 4} = \sqrt[4]{12x + 16}$

NAME ______________________________ DATE __________

Available as a transparency

Application Lesson Opener

For use with pages 445–452

The *mean*, or *average*, of a set of *n* numbers is the sum of the numbers divided by *n*.

To find the mean of the numbers 12, 14, 18, 13, and 14, first add the numbers. The sum is 71. Then divide by 5 (because there are 5 numbers): mean $= \frac{71}{5} = 14.2$.

Find the mean of the data set. If necessary, round your answer to the nearest hundredth.

1. Prices of used computer monitors at a store:
   55, 62, 59.95, 58.45, 68.90, 57.40, 49.75, 70, 60.75
2. Number of pages in books on a shelf:
   518, 144, 326, 803, 216, 84, 306
3. Number of Earth years for planets to orbit the sun:
   0.24, 0.62, 1, 1.88, 11.9, 29.5, 84, 165, 248
4. Ages of newly-elected United States Presidents at inauguration, 1945–1999:
   60, 62, 43, 55, 56, 61, 52, 69, 64, 46
5. Number of days in the months of the year:
   31, 28, 31, 30, 31, 30, 31, 31, 30, 31, 30, 31
6. Home runs by American League home run champions, 1923–1933:
   41, 46, 33, 47, 60, 54, 46, 49, 46, 58, 48
7. How are the two data sets in problems 1 and 3 similar? How are they different?

NAME ______________________________ DATE __________

Graphing Calculator Activity Keystrokes

For use with pages 453–454

TI-82

STAT [EDIT]1

Enter fat values in L1. STAT ▶ 1 ENTER

Use ▼ to cursor down to see the median.

2nd [STAT PLOT]1 ENTER

Choose the following. On; Type: (box plot)

Xlist: L1; Freq: 1

ZOOM 9 TRACE

STAT [EDIT]1

Enter caloric values in L2.

2nd [STAT PLOT]2 ENTER

Choose the following. On; Type: (histogram)

Xlist: L2; Freq: 1

ZOOM 9 TRACE

TI-83

STAT [EDIT]1

Enter fat values in L1. STAT ▶ 1 ENTER

Use ▼ to cursor down to see the median.

2nd [STAT PLOT]1 ENTER

Choose the following. On; Type: (box plot)

Xlist: L1; Freq: 1

ZOOM 9 TRACE

STAT [EDIT]1

Enter caloric values in L2.

2nd [STAT PLOT]2 ENTER

Choose the following. On; Type: (histogram)

Xlist: L2; Freq: 1

ZOOM 9 TRACE

Sharp EL-9600c

STAT [A] ENTER

Enter fat values in L1.

2ndF [QUIT] STAT [C] 1 ENTER

Use ▼ to cursor down to see the median.

2ndF [STAT PLOT][A] ENTER

Choose the following. on; DATA: X; ListX: L1;

Move cursor to GRAPH.

2ndF [STAT PLOT][E]1

ZOOM [A] 9 TRACE

STAT [A] ENTER

Enter caloric values in L2.

2ndF [STAT PLOT][B] ENTER

Choose the following. on; DATA: X; ListX: L2

Move cursor to GRAPH.

2ndF [STAT PLOT][A]1

ZOOM [A] 9 TRACE

Casio CFX-9850Ga Plus

From the main menu, choose STAT.

Enter fat values in List 1.

F2 F1

Use ▼ to cursor down to see the median.

SHIFT [SET UP] F2 EXIT F1 F6

Choose the following. Graph Type: MedBox; Xlist: List1; Freq: 1; Outliers: On.

EXIT SHIFT F3 0 EXE 50 EXE 1 EXE 0 EXE
1 EXE 1 EXE EXIT F1 F1 SHIFT F1

SHIFT [QUIT]

Enter caloric values in List 2.

F1 F6

Choose the following. Graph Type: Hist; Xlist: List 2; Freq: 1.

EXIT SHIFT F3 250 EXE 600 EXE 50 EXE
(-) 1 EXE 4 EXE 1 EXE EXIT F1 F1 F1

Choose: Start: 250; Pitch: 50 F6

SHIFT F1

NAME ______________________________ DATE __________

Practice A

For use with pages 445–452

Write the numbers in the data set in ascending order. Then find the mean, median, and mode of the data set.

1. 2, 3, 7, 1, 8, 7, 4, 5, 1, 8, 2, 6, 5, 9, 1

2. 10, 15, 8, 19, 12, 13, 10, 16, 12, 10

Find the range of the data set.

3. 18, 24, 37, 29, 13, 22, 25, 30

4. 123, 100, 132, 112, 148, 129, 138, 118

5. 3, 2, 1, 2, 3, 3, 1, 4

6. 105, 110, 104, 109, 110, 111, 108, 106

7. 2, 7, 150, 125, 3, 2, 1, 20

8. 88, 72, 84, 71, 73, 85, 90, 92

Find the lower and upper quartiles of the data set.

9. 5, 10, 7, 13, 12, 8, 15, 20, 10

10. 153, 146, 128, 144, 156, 120, 148, 160

11. 0, 3, 2, 4, 1, 6, 3, 5, 1

12. 38, 43, 32, 33, 37, 41, 44, 40, 38

Use the given information to draw a box-and-whisker plot of the data set.

13. minimum = 28
maximum = 54
median = 35
lower quartile = 32
upper quartile = 40

14. minimum = 120
maximum = 200
median = 160
lower quartile = 140
upper quartile = 185

Use the given intervals to make a frequency distribution of the data set.

15. Use five intervals beginning with 1–2.
1, 1, 1, 1, 2, 2, 2, 3, 3, 4, 4, 5, 5, 5, 6, 7, 7, 8, 8, 8, 8, 8, 9, 9, 10

16. Use five intervals beginning with 1–2.
1, 1, 2, 2, 1, 3, 2, 10, 4, 1, 6, 5, 3, 1, 9, 10, 6

Match the histograms with the data sets from Exercise 15 and Exercise 16.

17.

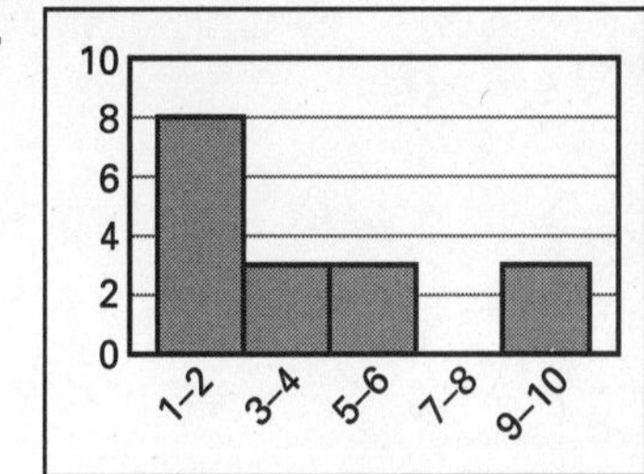

18.

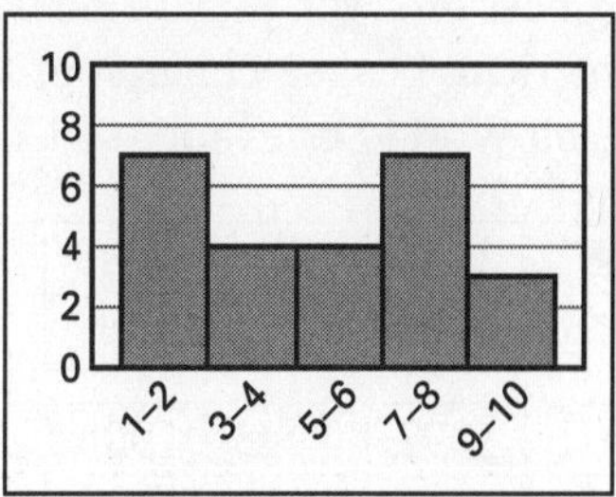

LESSON 7.7

NAME ______________________________ DATE __________

Practice B

For use with pages 445–452

Find the mean, median, and mode of the data set.

1. 6, 22, 4, 15, 10, 8, 8, 7, 14, 20
2. 10, 15, 12, 20, 25, 22, 28, 24, 22, 26
3. 53, 51, 47, 44, 60, 48, 44, 55, 44
4. 100, 150, 100, 120, 130, 125, 135, 140, 145

Find the range and standard deviation of the data set.

5. 47, 18, 65, 28, 43, 18
6. 35.8, 29.4, 32.1, 24.9, 30.5, 20.3

7. ***Reading Levels*** *The Pledge of Allegiance* contains 31 words. The bar graph at the right shows the number of words of different lengths in the pledge. Find the mean word length of the set of 31 words.

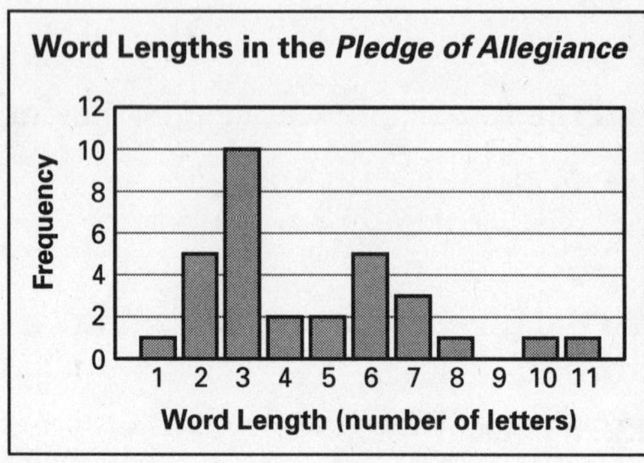

Lesson 7.7

Walking Shoes **In Exercises 8–10, use the following information.**

An important feature of walking shoes is their weight. The graph below shows the weight of the top-10 rated men's walking shoes.

8. Find the mean of the ten weights.
9. Find the median of the ten weights.
10. Find the mode of the ten weights.

Ranking	*Weight*	*Ranking*	*Weight*
1	24 oz	6	28 oz
2	22 oz	7	22 oz
3	26 oz	8	28 oz
4	28 oz	9	22 oz
5	24 oz	10	28 oz

World Series **In Exercises 11–13, use the following information.**

The World Series is a best-of-seven playoff between the National League champion and the American League champion. The table shows the number of games played in each World Series for 1981 through 1998.

Year	1981	1982	1983	1984	1985	1986	1987	1988	1989
Games	6	7	5	5	7	7	7	5	4

Year	1990	1991	1992	1993	1994	1995	1996	1997	1998
Games	4	7	6	6	0	6	6	7	4

11. Find the median of the number of games played.
12. Find the lower and upper quartiles of the number of games played.
13. Construct a box-and-whisker-plot of the number of games played.

LESSON **7.7**

NAME ______________________ DATE __________

Practice C

For use with pages 445–452

Find the range and the standard deviation of the data set.

1. 1, 4, 3, 2, 1, 2, 1, 3, 1
2. 6.5, 7.1, 6.8, 6.6, 6.8, 7.0
3. 105, 106, 104, 105, 107, 106
4. 20, 18, 36, 16, 16, 17, 21
5. *Test Scores* The bar graphs below represent three collections of test scores. Which collection has the smallest standard deviation?

A.

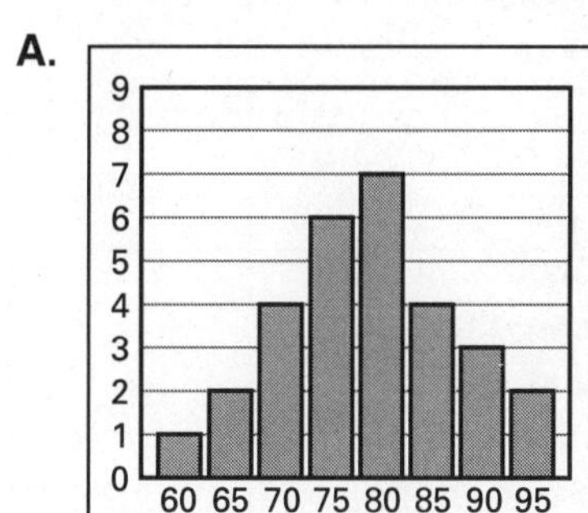

B.

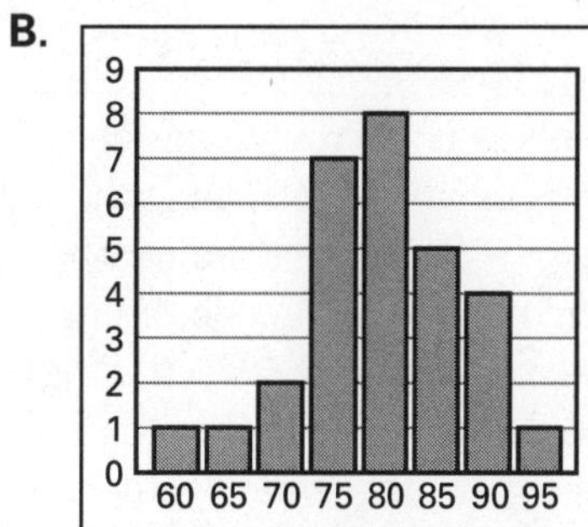

C.

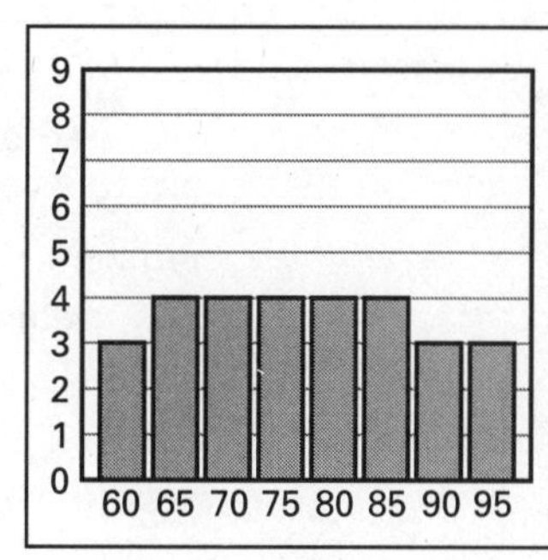

6. *Critical Thinking* Find the mean and median of the following data sets. When is the median a more accurate measure of central tendency?

a. 1, 1, 2, 3, 3, 2, 1, 50, 1, 3

b. 20, 25, 30, 24, 26, 1, 28, 25, 26, 28

Breakfast Cereals **In Exercises 7–9, use the following information.**
The number of calories in a 1-ounce serving of ten popular breakfast cereals is 116, 113, 104, 110, 119, 101, 106, 110, 106, 89.

7. Find the range of this data.
8. Find the mean of this data.
9. Find the standard deviation of this data. Round to three decimal places.

Manufacturing Couplers **In Exercises 10–12, use the following information.**
A company that manufactures hydraulic couplers takes ten samples from one machine and ten samples from another machine. The diameter of each sample is measured with a micrometer caliper. The company's goal is to produce couplers that have a diameter of exactly 1 inch. The results of the measurements are shown below.

Machine #1: 1.000, 1.002, 1.001, 1.000, 1.002, 0.999, 1.000, 1.002, 1.001, 1.001

Machine #2: 0.998, 0.999, 0.999, 1.000, 0.998, 0.999, 1.000, 1.000, 1.001, 0.999

10. Find the mean diameter for each machine.
11. Find the standard deviation for each machine.
12. Which machine produces the more consistent diameter?

History **In Exercises 13–15, use the following information.**
The table at the right gives the number of children of the Presidents of the United States.

Number of Children of U.S. Presidents
0, 0, 0, 0, 0, 0, 1, 1, 2, 2, 2, 2, 2, 2, 2, 2, 3, 3, 3, 3, 3, 3, 4, 4, 4, 4, 4, 4, 4, 5, 5, 5, 6, 6, 6, 6, 6, 7, 8, 10, 14

13. Make a frequency distribution of the data set using five intervals beginning with 0–2.
14. Draw a histogram of the data set.
15. Based on the histogram, which is the better measure of central tendency, the mean or median?

LESSON 7.7

NAME ______________________________ DATE ____________

Reteaching with Practice

For use with pages 445–452

GOAL Use measures of central tendency and measures of dispersion to describe data sets

VOCABULARY

The **mean**, or average, of n numbers is the sum of the numbers divided by n. The mean is denoted by $\bar{x}$ and is represented by $\dfrac{x_1 + x_2 + \cdots + x_n}{n}$.

The **median** of n numbers is the middle number when the numbers are written in order. (If n is even, the median is the mean of the two middle numbers.) The **mode** of n numbers is the number or numbers that occur most frequently. There may be one mode, no mode, or more than one mode. The **range** is the difference between the greatest and least data values. The **standard deviation** describes the typical difference (or deviation) between the mean and a data value, and is represented by

$$\sigma = \sqrt{\frac{(x_1 - \bar{x})^2 + (x_2 - \bar{x})^2 + \cdots + (x_n - \bar{x})^2}{n}}.$$

Lesson 7.7

EXAMPLE 1 *Finding Measures of Central Tendency*

Test Scores
32, 72, 81, 95, 98, 58, 77, 75, 83, 97, 45, 89, 93, 57, 82, 97, 52, 75

Find the mean, median, and mode of the data set listed above.

SOLUTION

To find the mean, divide the sum of the scores by the number of scores.

Mean:

$$\bar{x} = \frac{32 + 72 + 81 + 95 + 98 + 58 + 77 + 75 + 83 + 97 + 45 + 89 + 93 + 57 + 82 + 97 + 52 + 75}{18}$$

$$= \frac{1358}{18} \approx 75.4$$

To find the median, order the 18 numbers first. Because there is an even number of scores, the median is the average of the two middle scores.

32, 45, 52, 57, 58, 72, 75, 75, 77, 81, 82, 83, 89, 93, 95, 97, 97, 98

$$\text{Median} = \frac{77 + 81}{2} = 79$$

There are two modes, 75 and 97, because these numbers occur most frequently.

NAME ______________________ DATE __________

Reteaching with Practice

For use with pages 445–452

Exercises for Example 1

Find the mean, median, and mode of the data set.

1. 15, 11, 19, 15, 14, 14, 13, 17, 11, 12, 17, 15, 14, 15

2. 79, 78, 99, 98, 54, 75, 85, 61, 55, 86, 74

EXAMPLE 2 *Finding Measures of Dispersion*

Find the range and the standard deviation of the test scores from Example 1.

SOLUTION

To find the range, subtract the lowest score from the highest score.

Range $= 98 - 32 = 66$

To find the standard deviation, substitute the scores and the mean of 75.4 from Example 1 into the formula:

$$\sigma = \sqrt{\frac{(32 - 75.4)^2 + (45 - 75.4)^2 + (52 - 75.4)^2 + \cdots + (98 - 75.4)^2}{18}}$$

$$\approx \sqrt{\frac{6466}{18}} \approx \sqrt{359} \approx 18.9$$

Exercises for Example 2

3. Find the range and standard deviation of the data set in Exercise 1.

4. Find the range and standard deviation of the data set in Exercise 2.

LESSON 7.7

NAME ______________________________ DATE __________

Quick Catch-Up for Absent Students

For use with pages 445–454

The items checked below were covered in class on (date missed) ____________

Lesson 7.7: Statistics and Statistical Graphs

____ **Goal 1:** Use measures of central tendency and measures of dispersion to describe data sets. (pp. 445, 446)

Material Covered:

____ Example 1: Finding Measures of Central Tendency

____ Example 2: Finding Ranges of Data Sets

____ Example 3: Finding Standard Deviations of Data Sets

Vocabulary:

statistics, p. 445
measure of central tendency, p. 445
mean, p. 445
median, p. 445
mode, p. 445
measure of dispersion, p. 446
range, p. 446
standard deviation, p. 446

____ **Goal 2:** Use box-and-whisker plots and histograms to represent data graphically. (pp. 447, 448)

Material Covered:

____ Example 4: Drawing Box-and-Whisker Plots

____ Example 5: Making Frequency Distributions

____ Student Help: Skills Review

____ Example 6: Drawing Histograms

Vocabulary:

box-and-whisker plot, p. 447
lower quartile, p. 447
upper quartile, p. 447
histogram, p. 448
frequency, p. 448
frequency distribution, p. 448

Activity 7.7: Statistics and Statistical Graphs (p. 453, 454)

____ **Goal 1:** Use a graphing calculator to find statistics and draw statistical graphs.

____ Student Help: Keystroke Help

____ Student Help: Study Tip

____ Student Help: Study Tip

____ Other (specify) ____

____

Homework and Additional Learning Support

____ Textbook (specify) pp. 449–452

____

____ Internet: Extra Examples at www.mcdougallittell.com

____ *Reteaching with Practice* worksheet (specify exercises)____________

____ *Personal Student Tutor* for Lesson 7.7

Lesson 7.7

NAME ______________________________ DATE __________

Interdisciplinary Application

For use with pages 445–452

Populous U.S. Cities

GEOGRAPHY A city is a community where thousands or even millions of people live and work. Cities are the world's most crowded places. Cities offer many activities for residents and visitors. Art museums, musical performances, diverse restaurants, and large sports stadiums are a few of the activities of a large city.

People choose to live in or near cities for several reasons. The main reason is the number and variety of jobs available. Others may enjoy the rapid pace and bustling activity of city life.

In general, all communities in the United States with 2500 or more people, and smaller communities near big ones, are considered urban. Farms, and communities in uncrowded areas with fewer than 2500 people, are considered rural.

According to the United States Bureau of the Census, there are 219 cities in the United States that had 100,000 or more inhabitants in the year of 1996.

In Exercises 1–6, use the following information. Below is a list of the 35 largest cities according to population in the United States.

Population (in thousands) of the 35 largest cities
423, 430, 441, 441, 449, 470, 477, 480, 481, 498, 498, 511, 525, 541, 543, 558, 591, 597, 600, 657, 675, 680, 735, 747, 839, 1000, 1053, 1068, 1159, 1171, 1478, 1744, 2722, 3554, 7381

1. Find the mean, median, and mode of the data.
2. Find the range of the data.
3. Find the standard deviation of the data.
4. Draw a box-and-whisker plot of the data.
5. Make a frequency distribution of the data. Use fifteen intervals beginning with 1–500.
6. Draw a histogram of the data.

In Exercises 7 and 8, use the following information.

Here are the top 10 U.S. cities according to population in alphabetical order (not population order): Chicago, Dallas, Detroit, Houston, Los Angeles, New York, Philadelphia, Phoenix, San Antonio, and San Diego.

7. Using only your previous knowledge, place them in order from the most populated city to the least populated city.
8. Check your list by researching the populations of the cities. How many did you have in the right place?

NAME ______________________ DATE __________

Challenge: Skills and Applications

For use with pages 445–452

1. The z-score corresponding to a data point x is the number z defined by

$$z = \frac{x - \bar{x}}{\sigma},$$

where $\bar{x}$ is the mean of the data and σ is the standard deviation of the data.

a. Find the z-score of each data point in the data: 10, 35, 50, 60, 85.

b. What is the mean of the z-scores? What is their standard deviation? Do you think these values will be the same for any complete set of z-scores?

2. The *median-median line,* like the best-fitting line, is a linear function that approximates the relationship between two sets of data. You will find the equation of the median-median line for the following data: x: 3, 6, 8, 10, 12, 15, 19, 20, 22; y: 14, 16, 17, 22, 28, 28, 32, 37, 40.

a. Divide each set of data into 3 groups: the smallest third, the middle third, and the largest third (the data are already in order from smallest to largest), and find the 6 medians $x_1, x_2, x_3, y_1, y_2, y_3$, one for each of these groups.

b. Find an equation of the line passing through (x_1, y_1) and (x_3, y_3), $y = mx + b$. Use this to find the coordinates of the point P on this line for $x = x_2$.

c. Suppose the y-coordinate of P is y'. Let $w = \frac{y_2 + 2y'}{3}$. Find an equation of the line with slope m passing through $Q(x_2, w)$. This is the median-median line.

3. The number σ^2 (the standard deviation "without the radical") is called the *variance* of the data. Use the data x_1, x_2, x_3 and the following method to show how σ^2 can be written in simplest form.

a. Let $\bar{x}$ be the mean of the data and let $\overline{x^2}$ represent the mean of the *squares* of the data (i.e. the mean of the data set that consists of the squares of the given set). Show that

$\sigma^2 = \frac{1}{3}[x_1^2 + x_2^2 + x_3^2 - 2\bar{x} \cdot (x_1 + x_2 + x_3) + 3(\bar{x})^2]$.

b. Show that the right side of the equation in part (a) simplifies to $\overline{x^2} - (\bar{x})^2$.

CHAPTER 7

NAME ______________________________ DATE __________

Chapter Review Games and Activities

For use after Chapter 7

Solve the following problems and find the answer in the boxes at the bottom of the page. Place the letter from the box on the corresponding line with the number of the question. These letters will then give the answer to the following riddle: *How do you start a book about ducks?* Example: $f(x) = 3x - 2$ and $g(x) = x - 3$, find $f + g$ which is $4x - 5$, so C is placed on the line above the word "example."

1. Simplify $9^{-3/2}$.
2. Simplify $3^{-1/4}3^{9/4}$.
3. $f(x) = 3x - 2$, $g(x) = x - 3$ Find $f(g(x))$.
4. $f(x) = 2x - 9$, $g(x) = x - 2$ Find $f + g$.
5. Find $f^{-1}(x)$ of $f(x) = 2x^2 + 5$.
6. Solve $x^{3/2} + 2 = 29$.
7. $(3^3)^{1/3}(2^4)^{1/4}$

Use the following data set for 8–12.

1, 2, 3, 4, 4, 4, 5, 6, 7

8. Find the mean.
9. Find the median.
10. Find the standard deviation.
11. Find the mode.
12. Find the range.
13. $4^{3/2}$

S 1	M 27	A $\frac{1}{27}$	R $\pm\sqrt{\frac{x-5}{2}}$
T $3x - 11$	C $4x - 5$	E $\frac{2}{x^2 + 5}$	O 6
I 9	D 1.76	W 2	L 12
G 0	N 4	K 1.87	U 8

With ____ ____ ____ ____ ____ ____ ____-
1 8 2 9 3 5 7

____ ____ C ____ ____ ____ ____
10 13 example 4 6 12 11

Review and Assess

CHAPTER 7

NAME ______________________ DATE __________

Chapter Test A

For use after Chapter 7

Evaluate the expression without using a calculator.

1. $\sqrt[3]{-8}$
2. $25^{1/2}$
3. $27^{2/3}$
4. $8^{-1/3}$

Simplify the expression. Assume all variables are positive.

5. $(2^{1/3} \cdot 3^{1/3})^3$
6. $\sqrt[3]{8x^3y^6z^3}$
7. $\dfrac{x^3y^3}{(xy)^{-3}}$
8. $\sqrt{50} + \sqrt{8}$

Perform the indicated operation and state the domain. Let $f(x) = 3x$ and $g(x) = x - 5$.

9. $f(x) + g(x)$
10. $f(x) - g(x)$
11. $f(x) \cdot g(x)$
12. $\dfrac{f(x)}{g(x)}$
13. $f(g(x))$

Find the inverse function.

14. $f(x) = x + 9$
15. $f(x) = \frac{1}{2}x + 2$
16. $f(x) = 3x + 6$

Graph the function. Then state the domain and range.

17. $f(x) = \sqrt{x}$

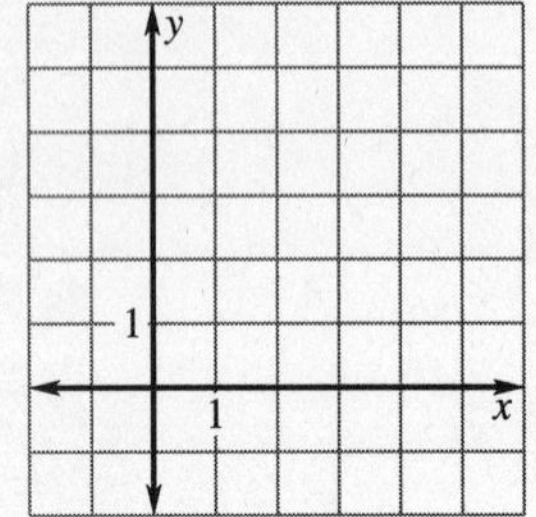

18. $f(x) = x^{1/3}$

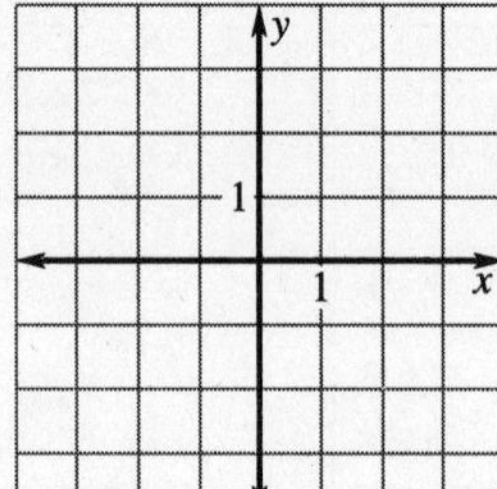

19. $g(x) = \sqrt{x - 3}$

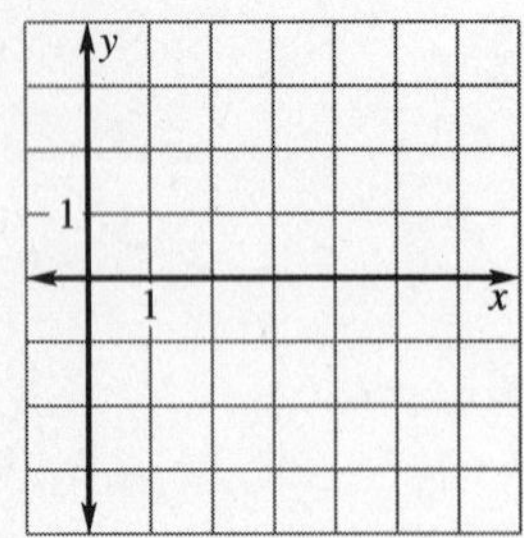

Answers

1. ____________
2. ____________
3. ____________
4. ____________
5. ____________
6. ____________
7. ____________
8. ____________
9. ____________
10. ____________
11. ____________
12. ____________
13. ____________
14. ____________
15. ____________
16. ____________
17. Use grid at left.
18. Use grid at left.
19. Use grid at left.

Review and Assess

CHAPTER 7 CONTINUED

NAME ______________________ DATE __________

Chapter Test A

For use after Chapter 7

Solve the equation. Check for extraneous solutions.

20. $x^{1/2} + 3 = 4$ **21.** $3\sqrt{2x + 4} = 12$ **22.** $\sqrt[3]{x^2 + 9} = 3$

Exam Scores In Exercises 23–25, suppose your exam scores on the ten exams taken in Algebra 2 are: 65, 75, 84, 72, 90, 92, 86, 95, 84, and 91.

23. Find the mean, median, mode, and range of the exam scores.

24. Draw a box-and-whisker plot of the exam scores.

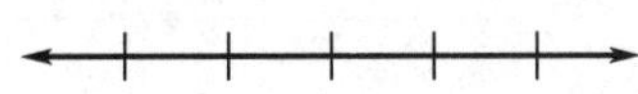

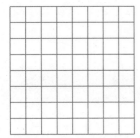

25. Make a frequency distribution using four intervals beginning with 60–69. Then draw a histogram of the data set.

20. ______________

21. ______________

22. ______________

23. ______________

____

____

____

24. Use space at left.

25. Use space at left.

Review and Assess

CHAPTER 7

NAME ______________________ DATE __________

Chapter Test B

For use after Chapter 7

Evaluate the expression without using a calculator.

1. $\sqrt[3]{-64}$ 2. $81^{1/2}$ 3. $(-27)^{2/3}$ 4. $125^{-1/3}$

Simplify the expression. Assume all variables are positive.

5. $(3^{1/3} \cdot 4^{1/3})^3$ 6. $\sqrt[3]{-8x^3y^3z^3}$ 7. $\left(\frac{4xy^{-1}}{16xy^2}\right)$ 8. $\sqrt{98} + \sqrt{2}$

Perform the indicated operation and state the domain. Let $f(x) = x - 1$ and $g(x) = 2x$.

9. $f(x) + g(x)$ 10. $f(x) - g(x)$ 11. $f(x) \cdot g(x)$

12. $\frac{f(x)}{g(x)}$ 13. $f(g(x))$

Find the inverse function.

14. $f(x) = 2x - 4$ 15. $f(x) = -\frac{2}{3}x + 4$ 16. $f(x) = x^2, x \ddagger 0$

Graph the function. Then state the domain and range.

17. $f(x) = \sqrt{x} + 1$

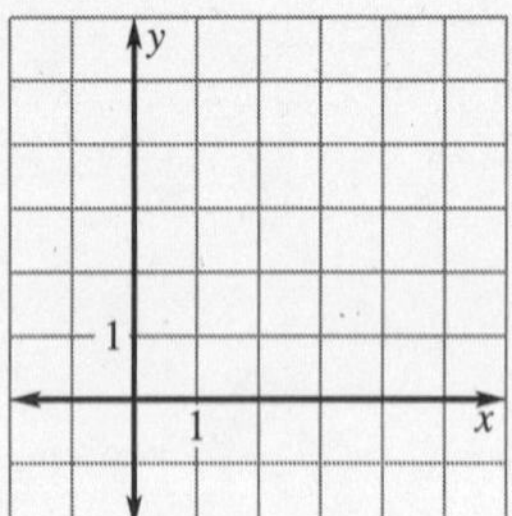

18. $f(x) = 2x^{1/3} - 3$

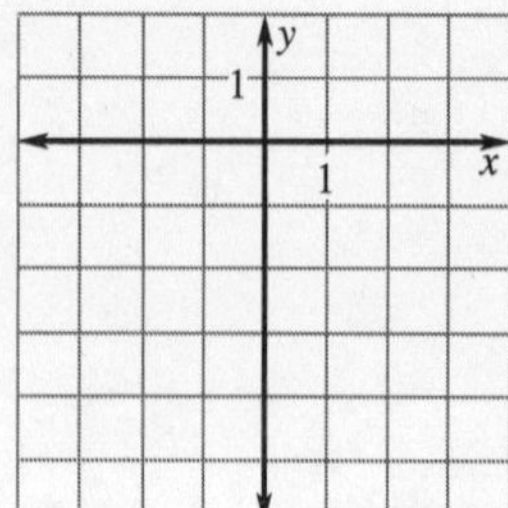

19. $f(x) = \sqrt{x - 7}$

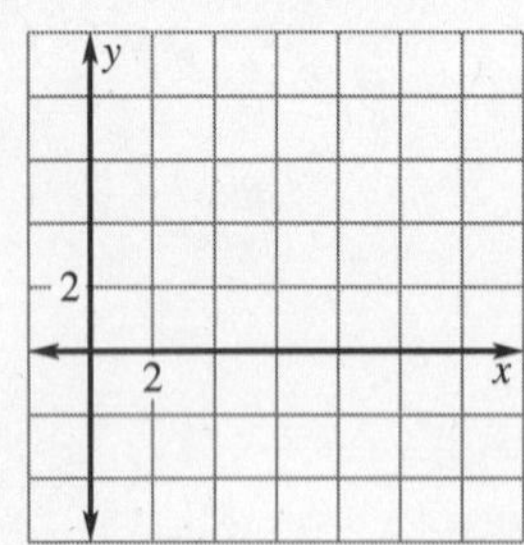

Answers

1. ______
2. ______
3. ______
4. ______
5. ______
6. ______
7. ______
8. ______
9. ______
10. ______
11. ______
12. ______
13. ______
14. ______
15. ______
16. ______
17. Use grid at left.
18. Use grid at left.
19. Use grid at left.

Review and Assess

NAME ______________________ DATE __________

Chapter Test B

For use after Chapter 7

Solve the equation. Check for extraneous solutions.

20. $\sqrt[3]{2x} = 4$

21. $\sqrt{3x} = \sqrt{x + 6}$

22. $\sqrt{x^2 + x - 3} = 3$

Polar Bears In Exercises 23–25, suppose a scientific team gathered the weights (in pounds) of ten polar bears. The weights are 964, 1002, 1026, 978, 1078, 925, 928, 1005, 964, and 894.

23. Find the mean, median, mode, range, and standard deviation of the weights.

24. Draw a box-and-whisker plot of the weights.

25. Make a frequency distribution using four intervals beginning with $875 - 929$. Then draw a histogram of the data set.

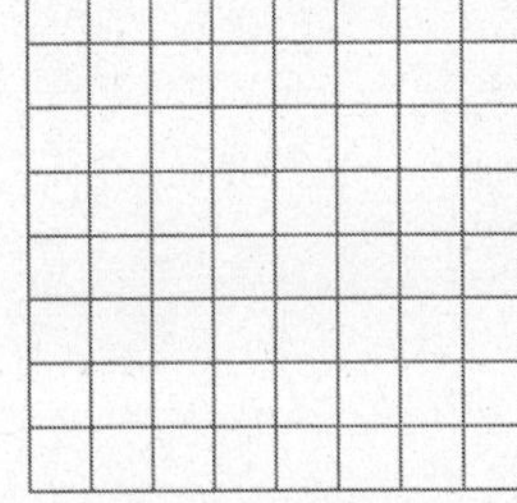

20. ______________

21. ______________

22. ______________

23. ______________

____

____

____

____

24. Use space at left.

25. Use space at left.

CHAPTER 7

NAME ______________________ DATE __________

Chapter Test C

For use after Chapter 7

Evaluate the expression without using a calculator.

1. $\sqrt[3]{-125}$
2. $27^{2/3}$
3. $\sqrt[4]{81}$
4. $\left(\frac{1}{216}\right)^{-1/3}$

Simplify the expression. Assume all variables are positive.

5. $(3^{1/2} \cdot 3^{1/3})$
6. $\sqrt[4]{32x^5y^4}$
7. $\left(\frac{27x^6}{8y^{12}}\right)^{2/3}$
8. $\sqrt[3]{54} + \sqrt[3]{2}$

Perform the indicated operation and state the domain. Let $f(x) = x - 1$ and $g(x) = x + 1$.

9. $f(x) + g(x)$
10. $f(x) - g(x)$
11. $f(x) \cdot g(x)$
12. $\frac{f(x)}{g(x)}$
13. $f(g(x))$

Find the inverse function.

14. $4x - 2y = 8$
15. $f(x) = x^2 + 5; x \geq 0$
16. $f(x) = (x - 7)^{1/3}$

Graph the function. Then state the domain and range.

17. $f(x) = 1 - \sqrt{x}$

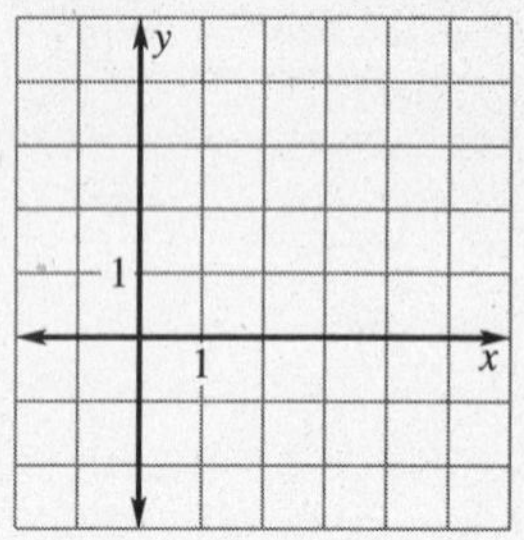

18. $f(x) = x^{1/2} - 2$

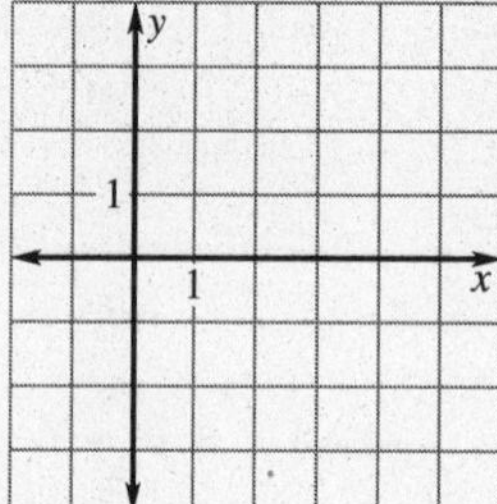

19. $f(x) = \sqrt[3]{x + 2} + 1$

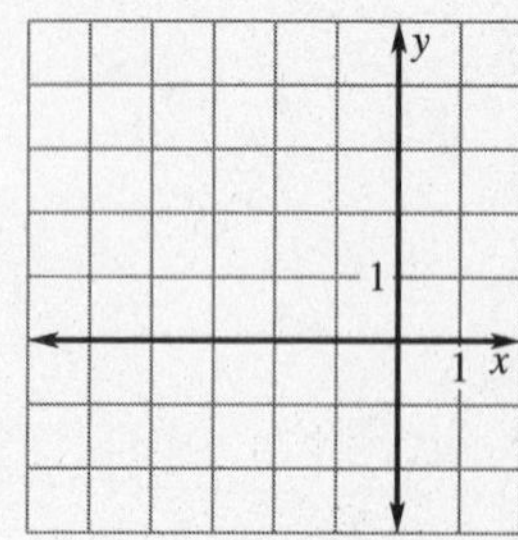

Answers

1. ____________
2. ____________
3. ____________
4. ____________
5. ____________
6. ____________
7. ____________
8. ____________
9. ____________
10. ____________
11. ____________
12. ____________
13. ____________
14. ____________
15. ____________
16. ____________
17. Use grid at left. ____________
18. Use grid at left. ____________
19. Use grid at left. ____________

Review and Assess

NAME ______________________ DATE __________

Chapter Test C

For use after Chapter 7

Solve the equation. Check for extraneous solutions.

20. $4 - x = \sqrt{10 - 3x}$

21. $5 = -\sqrt{7y - 3}$

22. $2(x + 2)^{1/3} = 6$

Basketball In Exercises 23–26, use the tables below which give the points scored in each game played by the boys and girls basketball teams this season.

Boys Team
56, 81, 80, 75, 48, 65, 90, 66, 70, 70

Girls Team
60, 72, 61, 58, 78, 65, 66, 55, 65, 73

23. Find the mean, median, mode, range, and standard deviation for each data set.

24. Interpret the data as to which team is more consistent in their scoring (use the standard deviation).

25. Draw a box-and-whisker plot of the *boys* points.

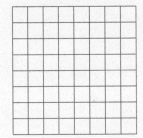

26. Make a frequency distribution of the *girls* points using five intervals beginning with 55–59. Then draw a histogram of this data.

20. ____________

21. ____________

22. ____________

23. ____________

24. ____________

25. Use space at left.

26. Use space at left.

NAME ______________________________ DATE __________

SAT/ACT Chapter Test

For use after Chapter 7

1. Evaluate $\sqrt[3]{-64}$.

   Ⓐ 8 Ⓑ -8
   Ⓒ 4 Ⓓ -4

2. Evaluate $\left(\frac{1}{125^{1/3}}\right)^{-1}$.

   Ⓐ -5 Ⓑ 5
   Ⓒ $\frac{1}{-5}$ Ⓓ $\frac{1}{5}$

3. What is the simplified form of $(3^{1/2} \cdot 8^{1/3})^2$?

   Ⓐ 4 Ⓑ 12
   Ⓒ $2\sqrt[3]{4}$ Ⓓ $24^{1/3}$

4. What is the simplified form of $\frac{1}{a^{-5/4}}$?

   Ⓐ $\frac{1}{a^{4/5}}$ Ⓑ $a^{4/5}$
   Ⓒ $a^{5/4}$ Ⓓ $a^{-5/4}$

In Exercises 5–7, perform the indicated operation. Let $f(x) = x + 1$ and $g(x) = x - 1$.

5. $f(x) + g(x)$

   Ⓐ $2x$ Ⓑ $x^2 - 1$
   Ⓒ $2x - 2$ Ⓓ $2x^2 - 1$

6. $f(x) \cdot g(x)$

   Ⓐ $2x^2 - 1$ Ⓑ $2x^2$
   Ⓒ $2x^2 + 1$ Ⓓ $x^2 - 1$

7. $f(g(x))$

   Ⓐ x Ⓑ $x^2 - 1$
   Ⓒ $x - 1$ Ⓓ $2x$

8. What is the solution of $2(x + 3)^{1/3} - 5 = 1$?

   Ⓐ $\frac{1}{24}$ Ⓑ -24
   Ⓒ 24 Ⓓ no solution

Quantitative Comparision In Exercises 9 and 10, choose the statement that is true about the given quantities.

Ⓐ The quantity in column A is greater.
Ⓑ The quantity in column B is greater.
Ⓒ The two quantities are equal.
Ⓓ The relationship cannot be determined from the given information.

9.

Column A	*Column B*
$x^{1/3}$	$y^{1/3}$

Ⓐ Ⓑ Ⓒ Ⓓ

10.

Column A	*Column B*
The *mean* of 1, 2, 3, 4, 5, 6, 7	The *median* of 1, 2, 3, 4, 5, 6, 7

Ⓐ Ⓑ Ⓒ Ⓓ

NAME ______________________ DATE __________

Alternative Assessment and Math Journal

For use after Chapter 7

JOURNAL 1. Your classmate, Jino, has simplified the following exponential and radical expressions and equations. He asks you to check his answers over before he submits them to the teacher. For each answer, determine if Jino's answer is correct or incorrect. If an answer is incorrect, find Jino's mistake. Give the correct answer and give a possible explanation for Jino's mistake.

a. $(27)^{1/3} = 9$

b. $(x^2)^3 = x^6$

c. $\sqrt{x^{16}} = x^4$

d. $\sqrt{81} = \sqrt{9} = 3$

e. $(4 - \sqrt{x})^2 = 16 - x$

f. When solving the equation $\sqrt{x+2} - 2 = x$, Jino's answer is $x = 3$ or $x = -2$.

MULTI-STEP PROBLEM 2. Given $f(x) = x - 3$ and $g(x) = 2x$ determine the following.

a. $f^{-1}(x)$

b. $g^{-1}(x)$

c. $f(g(x))$

d. $g(f(x))$

e. $[f(g(x))]^{-1}$, the inverse of $f(g(x))$ from (c).

f. $[g(f(x))]^{-1}$, the inverse of $g(f(x))$ from (d).

3. ***Critical Thinking*** Look at your answers to Exercise 2. Make a hypothesis about the inverse of a composition, $[f(g(x))]^{-1}$, and the inverses of its composing functions, $f^{-1}(x)$ and $g^{-1}(x)$. Does the same pattern seem to hold true for $[g(f(x))]^{-1}$? Write a general rule for finding the inverse of the composition of two functions.

CHAPTER 7 CONTINUED

Alternative Assessment Rubric

For use after Chapter 7

JOURNAL SOLUTION

1. **a–f.** Complete answers should include the following:

a. • Correct answer is $(27)^{\frac{1}{3}} = 3$.

• Explain that Jino divided 27 by 3 rather than finding the cube root of 27.

b. • Jino's answer is correct.

c. • Correct answer is $\sqrt{x^{16}} = x^8$

• Explain that Jino took the square root of 16 instead of multiplying the exponents $16 \cdot \frac{1}{2}$.

d. • Correct answer is $\sqrt{81} = 9$.

• Explain that Jino should take the square root of the value only once.

e. • Correct answer is $(4 - \sqrt{x})^2 = 16 - 8\sqrt{x} + x$.

• Explain that Jino multiplied incorrectly, forgetting the middle terms.

f. • Correct answer is $x = -2$ or $x = -1$.

• Explain that Jino did not isolate the radical expression before he squared both sides.

MULTI-STEP PROBLEM SOLUTION

2. **a.** $f^{-1}(x) = x + 3$ **b.** $g^{-1}(x) = \frac{x}{2}$

c. $f(g(x)) = 2x - 3$ **d.** $g(f(x)) = 2x - 6$

e. $[f(g(x))]^{-1} = \frac{x+3}{2}$ **f.** $[g(f(x))]^{-1} = \frac{x+6}{2} = \frac{x}{2} + 3$

3. ***Critical Thinking*** It seems that $[f(g(x))]^{-1} = g^{-1}(f^{-1}(x))$ and $[g(f(x))]^{-1} = f^{-1}(g^{-1}(x))$. This is logical because an inverse function performs the reverse operation as the original function. Therefore, the inverse of a composition of two functions performs the reverse operations in the reverse order.

MULTI-STEP PROBLEM RUBRIC

4 Answers to all parts are correct and explanations are thorough. Student is able to perform function compositions, inverses and operations.

3 Answers are correct and explanations show an understanding of functions. The process of finding compositions and inverses is correct.

2 Answers are partially correct. Explanations are incomplete or vague and show some understanding of compositions, inverses and operations, but application process is not always correct.

1 Answers are incomplete or incorrect. Explanations show little understanding of finding a composition of functions or finding the inverse of a function.

NAME ______________________________ DATE __________

Project: Walk This Way

For use with Chapter 7

OBJECTIVE **Examine the relationship between leg length and walking speed.**

MATERIALS tape measure, stopwatch or watch with second hand, and a straight, flat course (about 60 ft long)

INVESTIGATION Try walking across a flat, open space at an increasing speed. You'll notice that at some point you'll feel the urge to start running instead of walking. Your body is reaching its maximum walking speed. A person's leg length is related to his or her maximum walking speed.

1. Measure the length of the course in feet.
2. Measure each subject's leg length, from hip to heel in inches.
3. Instruct each subject to walk the course as fast as he or she can. But, no running is allowed! With each step, the rear foot must not leave the ground before the forward foot lands.
4. Time each subject as he or she walks the course.
5. Calculate each subject's speed (course length in feet divided by time in seconds).
6. Create a scatter plot of your data. Let the horizontal axis represent leg length, and let the vertical axis represent walking speed.
7. Athletic trainers use the formula $s = \frac{\sqrt{gl}}{12}$ to estimate a person's maximum walking speed. In the formula, s is the walking speed in feet per second, l is the leg length in inches, and g is the acceleration due to gravity, 386 in./s^2. Use the formula to calculate the predicted speed for each of your subjects. Then plot the curve $s = \frac{\sqrt{gl}}{12}$ on the same set of axes as your scatter plot.
8. Does the formula appear to predict the maximum walking speed accurately?

PRESENT YOUR RESULTS Write a report about your experiment. Include a statement of the goal of your project, and describe the procedures you followed. Then give your answers to the numbered questions above. Also describe any difficulties you had. For example, was it difficult to get precise measurements? How might you alter the experiment to improve the results?

Project: Teacher's Notes

For use with Chapter 7

GOALS
- Conduct an experiment to gather data.
- Graph and interpret a radical function to solve real-life problems.
- Compare experimental data to a theoretical model and analyze results.

MANAGING THE PROJECT You might find that it is most efficient to use the students in your class as the subjects for this experiment. It may be possible to devote some class time to collecting data for the entire class. However, for the data analysis portion of the experiment, students should work in pairs or small groups.

Encourage groups to make collective decisions and to prepare the final report jointly. If necessary, you can break the report into parts and require each student to write the first draft of one part.

To extend the project, you may wish to have students research the effects of gravity on other planets. How fast would you be able to walk on the moon or on Mars?

RUBRIC The following rubric can be used to assess student work.

4 The student conducts the experiment carefully and records accurate measurements. The report contains a detailed description of the methods used to gather data. The experimental data and theoretical results are presented in an easy-to-read format. The student writes reasonable explanations regarding any discrepancy between the experimental data and the theoretical results.

3 The student conducts the experiment carefully and records reasonably accurate measurements. The report contains a description of the methods used to gather data. The experimental data and theoretical results are presented. However, the student may not perform all calculations accurately or may not fully address discrepancies between the experimental data and the theoretical results.

2 The student attempts to conduct the experiment carefully and write a report summarizing the results. However, work may be incomplete or reflect misunderstandings. For example, the student may have measured some quantities using the wrong units. The report may indicate a limited grasp of certain ideas or may lack key supporting evidence.

1 The student leaves out portions of the report or does not show an understanding of key ideas. The student doesn't compare the experimental data with the theoretical results or fails to give reasonable explanations.

CHAPTER 7

NAME ______________________________ DATE __________

Cumulative Review

For use after Chapters 1–7

Identify the property shown. (1.1)

1. $-7 + 7 = 0$
2. $7 \cdot 9 = 9 \cdot 7$
3. $3 \cdot (5 \cdot 7) = (3 \cdot 5) \cdot 7$
4. $9(1) = 9$
5. $2(5 + 3) = 2 \cdot 5 + 2 \cdot 3$
6. $4 + 7 = 7 + 4$

Solve the equation. (1.3, 1.7)

7. $4x - 5 = 11$
8. $1.3x + 3.5 = 10$
9. $|x + 3| = 8$
10. $|10 - 4x| = 14$
11. $\frac{2}{3}x + \frac{5}{6} = \frac{1}{6}x + \frac{23}{6}$
12. $4(2x + 3) = 14$

Match the equation with the graph. (2.3, 2.8)

13. $y = \frac{1}{3}x + 1$
14. $y = x + \frac{1}{3}$
15. $y = |x - 3|$

A.

B.

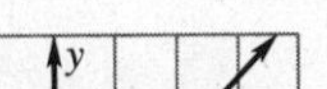

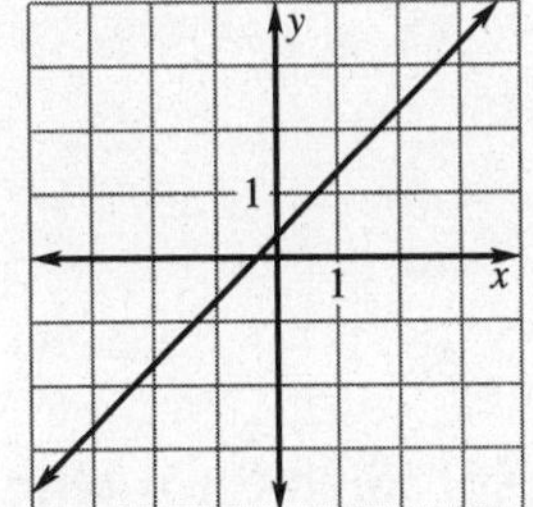

C.

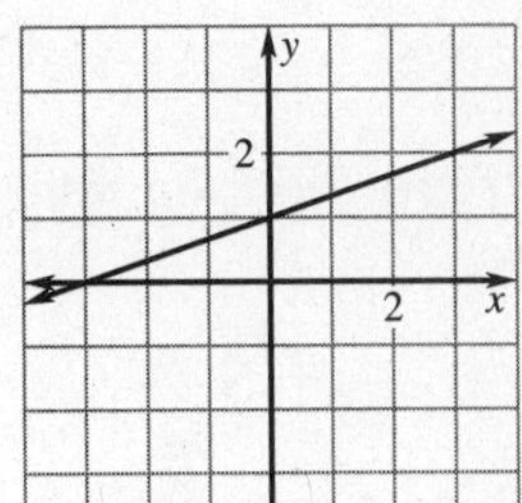

Draw a scatter plot of the data. Then state whether the data have a *positive correlation*, a *negative correlation*, or *relatively no correlation*. (2.5)

16.

x	1	1	2	4	5	6	8
y	1	2	3	5	5	7	9

17.

x	−2	−1	1	2	3	3	4	5
y	5	4	2	2	2	0	−1	−3

Find the minimum and maximum values of the objective function subject to the given constraints. (3.4)

18. Objective function:
$C = 2x + y$

Constraints:

$x \ge 0$
$y \ge 0$
$x + y \le 6$

19. Objective function:
$C = x + 3y$

Constraints:

$x \le 6$
$x \ge 1$
$y \le 5$
$y \ge 0$

20. Objective functions:
$C = 2x + 3y$

Constraints:

$x \ge -4$
$x \le 2$
$y \le 5$
$y \ge 1$

Perform the indicated matrix operation. If the operation is not defined, state the reason. (4.1, 4.2)

21. $\begin{bmatrix} -3 & -2 \\ 4 & 1 \end{bmatrix} + \begin{bmatrix} 4 & 3 \\ -2 & 0 \end{bmatrix}$

22. $\begin{bmatrix} 8 & -3 \\ 7 & -1 \end{bmatrix} - \begin{bmatrix} 4 & -5 \\ 0 & -3 \end{bmatrix}$

23. $-4\begin{bmatrix} 2 & 1 & 3 \\ 1 & -2 & -2 \end{bmatrix}$

24. $\begin{bmatrix} -3 & 4 \\ 2 & 1 \\ 4 & 0 \end{bmatrix}\begin{bmatrix} -1 & 2 \\ -1 & 3 \end{bmatrix}$

25. $\begin{bmatrix} 2 & 8 & 1 \\ 3 & 4 & -1 \end{bmatrix}\begin{bmatrix} 0 & 2 & 3 \\ 2 & -1 & 2 \end{bmatrix}$

26. $\begin{bmatrix} -\frac{1}{2} & \frac{1}{3} & -\frac{1}{2} \end{bmatrix}\begin{bmatrix} 6 \\ 0 \\ -4 \end{bmatrix}$

CHAPTER 7 CONTINUED

NAME ______________________ DATE __________

Cumulative Review

For use after Chapters 1–7

Evaluate the determinate of the matrix. (4.3)

27. $\begin{bmatrix} 1 & 2 \\ -2 & 4 \end{bmatrix}$

28. $\begin{bmatrix} 2 & 0 & 1 \\ 3 & 1 & -4 \\ -1 & 2 & -2 \end{bmatrix}$

29. $\begin{bmatrix} 1 & 4 & 5 \\ 2 & -2 & 0 \\ -2 & -1 & 4 \end{bmatrix}$

Find the inverse of the matrix. (4.4)

30. $\begin{bmatrix} 7 & -4 \\ -3 & 2 \end{bmatrix}$

31. $\begin{bmatrix} 8 & 1 \\ 2 & -1 \end{bmatrix}$

32. $\begin{bmatrix} 8 & -1 \\ 4 & -2 \end{bmatrix}$

Factor the expression. (5.2)

33. $4x^4 + 4x^2 + 1$

34. $a^4 - 8a^2 + 16$

35. $x^4 - y^4$

36. $8a^3 - 125$

37. $x^3 - x^2y - y^3 + xy^2$

38. $27x^3 - 8y^3$

Graph the inequality. (5.7)

39. $y \geq x^2 + 3$

40. $y \geq -x^2 - x + 5$

41. $y < 3x^2 - 12x + 11$

42. $y < (x + 2)^2$

43. $y \leq -(x - 3)^2 + 1$

44. $y \geq 4x^2 - 3$

Write a quadratic function in the specified form whose graph has the given characteristics. (5.8)

45. vertex form
vertex: $(3, -1)$
point on graph: $(4, 2)$

46. vertex form
vertex: $(-2, 0)$
point on graph: $(-3, 1)$

47. intercept form
x-intercepts: $-4, 2$
point on graph: $(3, -14)$

48. intercept form
x-intercepts: 2, 4
point on graph: $(1, -3)$

49. standard form
points on graph: $(0, -1), (3, 11), (-3, -1)$

Simplify the expression. (6.1)

50. $x^5 \cdot \frac{1}{x^2}$

51. $(x^3y^4)^{-3}$

52. $\frac{x^7}{x^{-3}}$

53. $(3x^2y^3)^{-2}$

54. $\frac{x^{-1}y^2}{x^{-4}y^{-3}}$

55. $-5x^{-4}y^0$

56. ***Planet Temperatures*** Pluto's surface temperature is believed to be −387°F, the lowest temperature observed on a natural body in our solar system. Measurements by the *Pioneer* probe indicate that Venus' surface temperature is 867°F. What is the difference between the two temperatures? **(1.1)**

57. ***Driving*** For a driver aged x years, a study found that a driver's reaction time $V(x)$ (in milliseconds) to a visual stimulus such as a traffic light can be modeled by: $V(x) = 0.005x^2 - 0.23x + 22$, $16 < x < 70$. At what age does a driver's reaction time tend to be greater than 20 milliseconds? **(5.7)**

ANSWERS

Chapter Support

Parent Guide

7.1: about 38.5 m^3 **7.2:** $y^{4/5}$
7.3: $g(f(x)) = 0.832x$; \$54.08
7.4: $f^{-1}(x) = \pm 2\sqrt{3 - x}$ **7.5:** 1.125 ft
7.6: 12 **7.7:** 10.8 points; about 6.08 points

Prerequisite Skills

1. $y = \frac{4x + 5}{3}$ **2.** $y = 6x + 12$
3. $y = -2x + 8$ **4.** $(x - 7)(x - 4)$
5. $(2x - 3)(x + 5)$ **6.** $(x - 8)(x + 3)$
7. $\frac{b^8}{a^6c^2}$ **8.** $\frac{1}{x^6y^9}$ **9.** $\frac{x^6}{25y^4}$ **10.** $2x^5 - 12x^4$
11. $16y^2 - 8y + 1$ **12.** $x^2 - 8x + 2$

Strategies for Reading Mathematics

1. 51; the box-and-whisker plot **2.** 8
3. a. 44% **b.** 6% **c.** 0%
4. Add the heights of the bars.

Lesson 7.1

Warm-up Exercises

1. 3 **2.** -11 **3.** 25 **4.** 7 or -7
5. 9 or -7

Daily Homework Quiz

1. $f(x) = \frac{1}{3}(x + 3)(x + 1)(x - 4)$
2.

$f(1)$	$f(2)$	$f(3)$	$f(4)$	$f(5)$
1	6	17	34	57
5	11	17	23	
6	6	6		

3. $f(x) = x^3 - 3x^2 + 4$

Lesson Opener

Allow 10 minutes.

1. 11 cm **2.** 8 ft **3.** 9 m **4.** 5 cm **5.** 1.2 in.
6. 3 m **7.** 6 cm **8.** 5 in. **9.** 7 mm **10.** 0.8 ft

Practice A

1. $11^{1/3}$ **2.** $5^{1/4}$ **3.** $23^{1/5}$ **4.** $7^{1/2}$ **5.** $17^{1/3}$
6. $2^{1/6}$ **7.** $8^{1/4}$ **8.** $15^{1/3}$ **9.** $10^{1/2}$ **10.** $3^{1/7}$
11. $6^{1/5}$ **12.** $21^{1/8}$ **13.** $\sqrt[3]{2}$ **14.** $\sqrt[4]{5}$
15. $\sqrt{11}$ **16.** $\sqrt[5]{6}$ **17.** $\sqrt[7]{23}$ **18.** $\sqrt[4]{31}$
19. $\sqrt{103}$ **20.** $\sqrt[3]{17}$ **21.** $\sqrt[3]{4}$ **22.** $\sqrt[8]{7}$
23. $\sqrt[5]{8}$ **24.** $\sqrt[14]{12}$ **25.** 2 **26.** 3 **27.** 2
28. 4 **29.** 1 **30.** 5 **31.** 1 **32.** 2 **33.** -2
34. 1.71 **35.** 2.88 **36.** 1.78 **37.** 1.32
38. 1.74 **39.** 1.52 **40.** 1.43 **41.** 2.29
42. 1.63 **43.** 1.32 **44.** 3.07 **45.** 2.24
46. 6 in. **47.** 8.08 cm.

Practice B

1. $7^{1/3}$ **2.** $5^{2/3}$ **3.** $11^{5/2}$ **4.** $12^{5/3}$ **5.** $15^{7/3}$
6. $(-9)^{5/3}$ **7.** $(-42)^{2/7}$ **8.** $(-10)^{8/3}$
9. $\sqrt[3]{19}$ **10.** $\sqrt[5]{43}$ **11.** $(\sqrt[3]{6})^2$ **12.** $(\sqrt[3]{9})^4$
13. $(\sqrt[4]{8})^3$ **14.** $(\sqrt[3]{-6})^2$ **15.** $(\sqrt[3]{-10})^4$
16. $(\sqrt[7]{-14})^3$ **17.** 16 **18.** 216 **19.** 8
20. 729 **21.** 16 **22.** 4 **23.** 32 **24.** -4
25. -32 **26.** 2.65 **27.** 3.00 **28.** -2.61
29. 2.93 **30.** -2.47 **31.** 2.21 **32.** 25.92
33. 4148.54 **34.** 291,461.63 **35.** 4.50 cm
36. -12, 12 **37.** 9.00 **38.** 16
39. 511.48 cm^3 **40.** 8 cm **41.** 556.28 cm^3
42. 8.22 cm

Practice C

1. 27 **2.** 729 **3.** 9 **4.** $\frac{1}{2}$ **5.** $\frac{1}{125}$ **6.** $\frac{1}{100{,}000}$
7. $\frac{1}{64}$ **8.** 256 **9.** $-\frac{1}{8}$ **10.** 3596.65
11. -106.17 **12.** 0.03 **13.** 0.15
14. 2002.65 **15.** 6.85 **16.** $-13{,}593.93$
17. 15.00 **18.** 0.10 **19.** 2.68 **20.** 1, -1
21. -1 **22.** 4.61, 2.39 **23.** -2.92
24. -0.99 **25.** 0.67, -1.67 **26.** 1.41, -1.41
27. -1.33 **28.** = **29.** > **30.** < **31.** =
32. $\approx$ 1.4 in. **33.** $\sqrt[n]{a^n} \neq a$ when $a < 0$ and n is even.

Lesson 7.1 *continued*

Reteaching with Practice

1. ± 4 **2.** -1 **3.** none **4.** 5 **5.** 0 **6.** ± 7 **7.** 9 **8.** 7 **9.** $\frac{1}{27}$ **10.** 2 **11.** -2 **12.** 216 **13.** ± 3.98 **14.** -12.83 **15.** 214.25 **16.** 19,600 **17.** ± 3.55 **18.** 5 **19.** ± 3.05 **20.** 3.58 **21.** 2.64

Interdisciplinary Application

1.

Planet	*a*
Mercury	0.387
Venus	0.723
Earth	1.000
Mars	1.524
Jupiter	5.201
Saturn	9.538
Uranus	19.181
Neptune	30.057
Pluto	39.510

2. Between 2 and 3.5 astronomical units

3. Between Mars and Jupiter

Challenge: Skills and Applications

1. $\frac{\sqrt[4]{125}}{5}$ **2.** $\frac{\sqrt[3]{4}}{2}$ **3.** $3\sqrt[5]{3}$

4. a. 1.91, 1.63, 1.21, 1.04, 1.02 **b.** 7^x approaches 1 **c.** *Sample answer:* $x^{1/x}$ approaches 1; 1.38, 1.26, 1.16, 1.08, 1.05, 1.007 **d.** *Sample answer:* x^x approaches 1; 0.72, 0.79, 0.86, 0.92, 0.95, 0.99 **5.** 9 **6.** 64 **7.** $\sqrt[11]{729}$ **8.** 1, 32 **9.** $\frac{25}{4}$ **10.** $\frac{1}{27}, -1$

11. $\sqrt[m]{x^n} = (x^n)^{1/m} = x^{n/m} = (x^{1/m})^n = \left(\sqrt[m]{x}\right)^n$

Lesson 7.2

Warm-up Exercises

1. 1024 **2.** 64 **3.** 9 **4.** $\frac{1}{8}$ **5.** 1296

Daily Homework Quiz

1. $2\sqrt{6}$ **2.** $9 \cdot 3^{1/2}$ **3.** $\pm\frac{1}{2}$ **4.** 6.08 **5.** 1.86 **6.** ± 2.05 **7.** 5.75

Lesson Opener

Allow 20 minutes.

Column 1: $n^8, 64d^3, j^5, \frac{1}{k^4}$

Column 2: $\frac{4}{u^7}, a^4, 243h^5, \frac{125}{z^6}$

Column 3: $32x^5, \frac{2}{c^8}, u^6, \frac{1}{t^2}$

Column 4: $100m^2, x^8, \frac{1}{g^4}, \frac{u^{12}}{2401}$

Column 5: $\frac{1}{z^6}, \frac{1}{y^3}, v^{10}, p^2$

Column 6: $t^6, e^{11}, 343q^3, \frac{1}{x^4}$

Practice A

1. $4^{5/3}$ **2.** $6^{1/4}$ **3.** $5^{2/3} \cdot 3^{2/3}$ **4.** $\frac{1}{13^{5/4}}$ **5.** $10^{1/6}$ **6.** $\frac{2^{1/8}}{3^{1/8}}$ **7.** $\sqrt{21}$ **8.** $\sqrt[6]{5}$ **9.** 2 **10.** $\frac{1}{5}$ **11.** $\sqrt[10]{3}$ **12.** $3\sqrt{2}$ **13.** $x^{3/4}$ **14.** $x^{8/3}$ **15.** $x^{1/10}$ **16.** $\sqrt{3}\,x^{1/2}$ **17.** $3x^{1/3}$ **18.** $\frac{1}{x^{7/2}}$ **19.** x **20.** $\frac{1}{x^2}$ **21.** $\frac{10}{x^{1/2}}$ **22.** $6\sqrt{3}$ **23.** $2\sqrt{7}$ **24.** $15\sqrt[5]{22}$ **25.** $-5\sqrt{x}$ **26.** $-3\sqrt[3]{x}$ **27.** $2\sqrt[4]{x}$ **28.** $7x\sqrt{x}$ **29.** $\frac{x}{y\sqrt{y}} = \frac{x\sqrt{y}}{y^2}$ **30.** $yz^2\sqrt{x^3y}$ **31.** $3x\sqrt[3]{yz}$ **32.** $6x^2\sqrt{x}$ **33.** $\frac{y^2\sqrt{xz}}{x^2}$ **34.** $2\sqrt{2}$ in.

Practice B

1. $5^2 = 25$ **2.** $3^{-1/2}$ **3.** $7^{5/3}$ **4.** $12^{1/4}$ **5.** $\sqrt[3]{8} = 2$ **6.** 2 **7.** $3^{-2/3}$ **8.** $\frac{4}{5}$ **9.** $10^{1/3}$ **10.** $3x$ **11.** $\sqrt[3]{2}\,x$ **12.** x **13.** $\frac{x^{1/2}}{2}$ **14.** $2x^{1/4}$ **15.** $3x$ **16.** $2x$ **17.** $x^{2/3}$ **18.** $4y\sqrt[4]{x}$ **19.** $x^{3+\sqrt{5}}$ **20.** x^3 **21.** $4^{\sqrt{2}}x^{\sqrt{2}}$ **22.** $x^{-4\sqrt{3}} = \frac{1}{x^{4\sqrt{3}}}$ **23.** $\frac{1}{x^{\sqrt{2}}}$ **24.** $x^{\sqrt{6}}$ **25.** $3\sqrt[3]{3}$ **26.** $-\sqrt[4]{15}$ **27.** $8(2^{1/3})$ **28.** $2\sqrt{2}$

Lesson 7.2 *continued*

29. $3\sqrt[3]{5}$ **30.** $-2\sqrt[5]{3}$
31. diameter = 5.88×10^{17} miles; thickness = 5.88×10^{16} miles **32.** 12.5 in.
33. 2.5 in. **34.** 0.2

Practice C

1. $5 \cdot 3^{3/4}$ **2.** $2^{13/24}$ **3.** $5^{4/15}$ **4.** $6^{1/10}$
5. $\frac{1}{8}$ **6.** $\sqrt[24]{2}$ **7.** $\sqrt{2}$ **8.** 4 **9.** $\frac{1}{\sqrt{10}} = \frac{\sqrt{10}}{10}$
10. $\frac{x^{1/4}}{y^{1/3}}$ **11.** $\frac{9y^{1/3}z^2}{4x^{3/2}}$ **12.** $\frac{2z^{1/8}}{x^{1/3}y^{5/4}}$ **13.** $3x^{1/2}$
14. $4x^4$ **15.** $\sqrt[4]{x}$ **16.** $\sqrt[4]{6}$ **17.** $5x^2\sqrt[3]{y^2z}$
18. $\sqrt[10]{x}$ **19.** $\frac{2\sqrt[3]{y^2}}{y}$ **20.** $-\sqrt[5]{y}$
21. $-15x^2y\sqrt{y}$ **22.** 2.7×10^{12} meters
23. 5.3×10^{12} meters

Reteaching with Practice

1. $x^{3/2}$ **2.** $\frac{1}{y^{2/3}}$ **3.** 256 **4.** $y^{1/3}$ **5.** 4
6. $z^{7/6}$ **7.** 4 **8.** 6 **9.** 2 **10.** 5 **11.** $2\sqrt[3]{4}$
12. $\frac{\sqrt[4]{18}}{3}$ **13.** $4x^2\sqrt[4]{y}$ **14.** $\frac{2x\sqrt{y}}{3z}$ **15.** $\sqrt[5]{3}$
16. $11(2^{1/8})$ **17.** $6\sqrt{x}$

Real-Life Application

1. 8306 days **2.** 400,099 hours **3.** about 2.6 bulbs **4.** about 4 bulbs

Challenge: Skills and Applications

1. $\sqrt[3]{3}$ **2.** 2 **3.** $\sqrt[3]{10}$ **4.** $\sqrt{2}$ **5.** $\sqrt[6]{5}$
6. $3\sqrt[3]{9}$ **7.** -2 **8.** -1 **9.** $-3, 2$
10. a. $1 + (m + n)x$ **b.** $1 + (m + n)x + mnx^2$
c. The term mnx^2 is much smaller than either of the other terms because it is a multiple of $x^2 \approx 0.0001$, if $x \approx 0.01$.

11. a. Write f_{n-1} as $\frac{w}{2}$. Then

$$f_n = \sqrt{\frac{1 + \frac{w}{2}}{2}} = \frac{\sqrt{2 + w}}{2};$$

$$\frac{\sqrt{2 + \sqrt{2 + \sqrt{2 + \sqrt{2}}}}}{2}$$ **b.** 3.14

Quiz 1

1. 8 **2.** $\frac{1}{125}$ **3.** -16 **4.** 4 **5.** 2.45, -2.45
6. -0.59 **7.** $3^{\frac{1}{5}}$ **8.** $\frac{2\sqrt[3]{9}}{3}$ **9.** $3\sqrt[3]{3}$
10. $4(2^{\frac{2}{3}})$ **11.** $x^{\frac{7}{6}}$ **12.** $x^{\frac{1}{2}}$ **13.** $3xy\sqrt[3]{3x^2}$
14. $x^{\frac{8}{3}}y^{\frac{1}{4}}$ **15.** 2.4 m

Lesson 7.3

Warm-up Exercises

1. $4x^2 + 4$ **2.** $x^2 + x - 1$ **3.** $-x^2 + x - 3$
4. $x^2 - x - 2$ **5.** $x^2 - 4x + 4$

Daily Homework Quiz

1. 6 **2.** $\sqrt{17}$ **3.** 5 **4.** $y^{11/6}$
5. $11z^2\sqrt{z} - 14$ **6.** $5\sqrt[3]{4}$

Lesson Opener

Allow 15 minutes.

1. 42 **2.** $14t$ **3.** $42s$ **4.** $28x + 70$ **5.** 6
6. $\frac{300}{r}$ **7.** $\frac{15}{p}$ **8.** $\frac{300}{75 - x}$ **9.** 20
10. $-16x^2 + 48x$ **11.** $-144x^2 + 144x$
12. $-16t^2 + 16t + 32$

Graphing Calculator Activity

1. $8x - 5$ **2.** $-2x + 9$ **3.** $15x^2 - 11x - 14$

Practice A

1. $3x + 1$ **2.** $x^2 + 2x + 2$ **3.** $2x^2 - 2x + 2$
4. $7x^{1/2}$ **5.** $x - 3$ **6.** $-x + 2$ **7.** $x^2 - x - 2$
8. $-x^{3/2}$ **9.** $6x - 3$ **10.** $3x^2 + x - 2$
11. $2x^3 + 2x^2 - 2x$ **12.** $6x$ **13.** $\frac{3x}{x + 2}$
14. $\frac{x^2 + 1}{x - 2}$ **15.** $\frac{x - 2}{x^2 + x - 4}$ **16.** $\frac{x^{1/6}}{2}$
17. $2x + 10$ **18.** $\sqrt{4x + 9}$ **19.** $x^2 - 2x + 3$
20. $x^{3/20}$ **21.** All real numbers **22.** All real numbers **23.** All real numbers **24.** All real numbers except $x = 3$ **25.** All real numbers

Lesson 7.3 *continued*

26. All real numbers **27.** All real numbers **28.** All real numbers except $x = 0$ **29.** All real numbers **30.** $P(x) = 0.75x - 20{,}000$; \$730,000

Practice B

1. $3x^3 - x^2 + 12x - 2$; $3x^3 - 3x^2 - 2x$
2. $7x^{2/3}$; $x^{2/3}$ **3.** $2x^3 + x^2 + 2x + 3$; $2x^3 - x^2 - 8x + 5$ **4.** $\frac{5}{8}x^{3/4}$; $\frac{3}{8}x^{3/4}$
5. $-x^3 + x^2 + 4x + 2$
6. $x^6 + 3x^4 + 3x^3 + 2x^2 + 9x + 6$
7. $4x^{7/12}$ **8.** $8x^{-1/2} = \dfrac{8}{x^{1/2}}$ **9.** $\dfrac{3x^2 - x + 1}{x + 3}$
10. $\dfrac{3x + 5}{2x^2 - 1}$ **11.** $2x^{5/3}$ **12.** $\dfrac{3^{1/4}}{x}$
13. $f(g(x)) = 6x + 3$, $g(f(x)) = 6x + 1$
14. $f(g(x)) = x^2 - 4x + 5$, $g(f(x)) = x^2 - 1$
15. $f(g(x)) = -(x + 4)^{1/2}$, $g(f(x)) = -x^{1/2} + 4$
16. $f(g(x)) = 3x^{2/5}$, $g(f(x)) = \sqrt{3}\,x^{2/5}$
17. $4x^{1/2} + x + 3$; nonnegative real numbers
18. $x + 3 - 4x^{1/2}$; nonnegative real numbers
19. $4x^{3/2} + 12x^{1/2}$; nonnegative real numbers
20. $\dfrac{x + 3}{4x^{1/2}}$; positive real numbers
21. $4(x + 3)^{1/2}$; real numbers greater than or equal to -3.
22. $4x^{1/2} + 3$; nonnegative real numbers
23. $f(x) = x - 100$ **24.** $g(x) = 0.75x$
25. $g(f(x)) = 0.75x - 75$
26. $f(g(x)) = 0.75x - 100$ **27.** Discount

Practice C

1. $x^5 - x^3 + 3x^2 + 6x - 9$; $-x^5 + 3x^3 + 3x^2 - 2x + 7$ **2.** $10x^{2/5} - 2x^{-1}$; $2x^{2/5} + 8x^{-1}$ **3.** $x^2 + 10x + 1$; $x^2 - 4x - 3$
4. $5x^{1/6} - 7x^3 - 1$; $x^{1/6} + 3x^3 - 1$
5. $x^5 + 4x^4 - x^3 - 15x^2 - 16x + 30$
6. $5x^{5/8} - 5x^{1/4} - 3x^{3/8} + 3$
7. $3x + x^{1/3}$ **8.** $\dfrac{16}{x^{7/3}}$
9. $f(g(x)) = (x^2 + 4)^{1/2}$; $g(f(x)) = x + 4$
10. $f(g(x)) = \dfrac{1}{(3x - 1)^2}$; $g(f(x)) = \dfrac{3}{x^2} - 1$
11. $f(g(x)) = (2x)^{3/4}$; $g(f(x)) = 2x^{3/4}$
12. $f(g(x)) = \dfrac{3}{2x^{1/2}}$; $g(f(x)) = \dfrac{2\sqrt{3}}{x^{1/2}}$
13. $f(g(x)) = \dfrac{1}{\sqrt{x^2 + 2x}}$; all real numbers less than -2 or greater than 0
14. $g(f(x)) = \dfrac{1 + 2x^{1/2}}{x}$; positive real numbers
15. $\dfrac{f(x)}{g(x)} = \dfrac{1}{x^{5/2} + 2x^{3/2}}$; positive real numbers
16. $\dfrac{g(x)}{f(x)} = x^{5/2} + 2x^{3/2}$; nonnegative real numbers
17. $f(f(x)) = x^{1/4}$; nonnegative real numbers
18. $g(g(x)) = x^4 + 4x^3 + 6x^2 + 4x$; all real numbers **19.** True **20.** False; Examples vary.
21. True **22.** False; Examples vary.
23. False; Examples vary. **24.** False; Examples vary.
25. *Sample answer:* $f(x) = \sqrt{x}$, $g(x) = 2x + 1$
26. *Sample answer:* $f(x) = \dfrac{1}{x}$, $g(x) = 3x + 2$
27. Let $f(x) = 0.6x$, $g(x) = x - 5$, $h(x) = 0.9x$
$f(g(h(x))) = 0.54x - 3$
$f(h(g(x))) = 0.54x - 2.7$
$g(f(h(x))) = 0.54x - 5$
$g(h(f(x))) = 0.54x - 5$
$h(f(g(x))) = 0.54x - 2.7$
$h(g(f(x))) = 0.54x - 4.5$; First the store will deduct the \$5 coupon. Then it makes no difference in what order they take the 40% and 10% discount.

Reteaching with Practice

1. $2 + 2x$; all real numbers **2.** $2 - 4x$; all real numbers **3.** $4x - 2$; all real numbers
4. $6x$; all real numbers **5.** $3x^{7/2}$; all nonnegative real numbers **6.** $2x^3 + 6x^2$; all real numbers
7. $\dfrac{2}{x^{1/3}}$; all real numbers except 0
8. $\dfrac{-7x + 1}{x}$; all real numbers except 0
9. $\dfrac{2}{x - 2}$; all real numbers except 2

Lesson 7.3 *continued*

10. $\frac{2}{x} - 2$; all real numbers except 0

11. x; all real numbers except 0

12. $x - 4$; all real numbers

Interdisciplinary Application

1. $A(r(t)) = \pi(0.6t)^2 = 0.36\pi t^2$ **2.** input: time t; output: area A **3.** 4.5 ft^2 **4.** 313.91 ft^2 **5.** 20 ft **6.** 9.5 seconds

Challenge: Skills and Applications

1. i, ii, iv **2.** i, iii, vi **3.** v **4.** none

5. abx^{m+n} **6.** $ab^m x^{mn}$ **7.** $a^n bx^{mn}$ **8.** $x^{n/2}$

9. $3^n x^{2n}$ **10.** $\frac{2^n}{x^n}$ **11.** no; $M + N \geq P$ (the maximums may occur at different values of x).

Lesson 7.4

Warm-up Exercises

1. $y = 2x$ **2.** $y = 2x + 1$ **3.** $y = -3x + 6$

4. $y = -\frac{1}{3}x + 1$ **5.** $y = \frac{3}{4}x + 2$

Daily Homework Quiz

1. $x^3 + x^2 + 4x - 2$; all real numbers

2. $x^5 + 2x^3 - 8x$; all real numbers

3. $x^6 + 8x^4 + 16x^2 - 2$; all real numbers

4. $\frac{x^3 + 4x}{x^2 - 2}$; all real numbers except $\pm\sqrt{2}$

Lesson Opener

Allow 10 minutes.

1–6. WHEEEEEEEEEE **7.** *Sample answer:* In each of problems 2–6, the functions "undo" one another, so each composition of functions gives x.

Practice A

1.

x	3	5	7	9	11
y	−2	−1	0	1	2

2.

x	1	−2	4	−1	0
y	0	1	2	3	4

3. no **4.** yes **5.** yes **6.** B **7.** A **8.** C

9–12. Show $f(g(x)) = x$ and $g(f(x)) = x$.

13. $i = \frac{C}{2.54}$; 16.54 in. **14.** $r = \frac{C}{2\pi}$; 4.46 in.

Practice B

1.

x	−6	−3	0	3	6
y	1	2	3	4	5

2.

x	1	2	4	6	0
y	−1	$-\frac{2}{3}$	0	$\frac{1}{2}$	3

3. yes **4.** no **5.** no **6.** no **7.** yes **8.** yes

9–16. Show $f(g(x)) = x$ and $g(f(x)) = x$.

17. $y = \frac{1}{4}x$ **18.** $y = -x + 5$ **19.** $y = \frac{1}{3}x - \frac{1}{3}$

20. $y = \frac{1}{4}x + \frac{9}{4}$ **21.** $y = 2x - 12$

22. $y = \frac{3}{2} - \frac{1}{2}x$

23. $y = -\sqrt{x - 3}, y = \sqrt{x - 3}$

24. $y = -\sqrt{x + 1}, y = \sqrt{x + 1}$

25. $y = x^2, x \geq 0$

26. [graph] **27.** [graph]

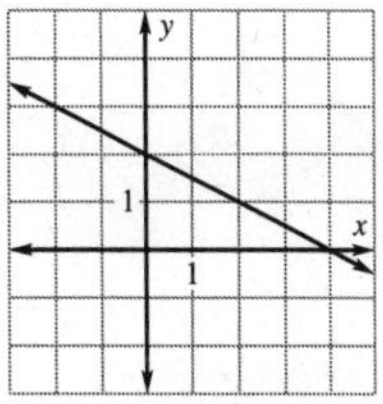

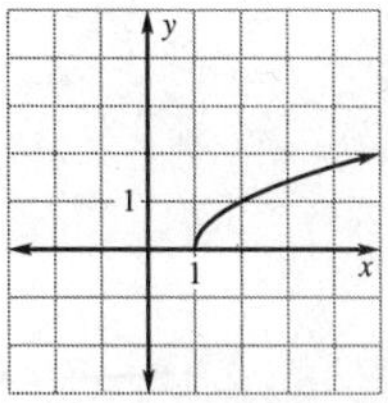

28. [graph] **29.** $C = K - 273.15$, 21.85°C

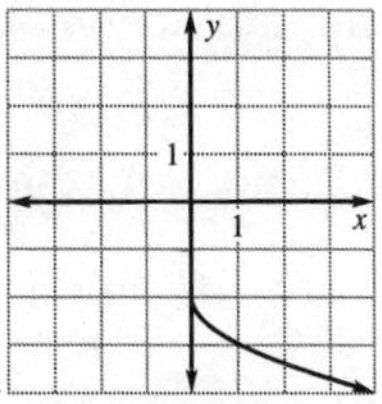

30. $R = \frac{S}{0.75}$, \$26.51

Practice C

1–6. Show $f(g(x)) = x$ and $g(f(x)) = x$.

7. $f^{-1}(x) = \frac{1}{4} - \frac{1}{4}x$ **8.** $f^{-1}(x) = \frac{1}{3}x - \frac{8}{3}$

9. $f^{-1}(x) = x^2 - 1, x \geq 0$

10. $f^{-1}(x) = \frac{1}{2}x^2 + \frac{3}{2}, x \geq 0$

11. $f^{-1}(x) = 4 - x^2, x \geq 0$

12. $f^{-1}(x) = \frac{1}{5}x^3 + \frac{3}{5}$

13. $f^{-1}(x) = \sqrt{x - 7}; x \geq 7$

14. $f^{-1}(x) = \sqrt[3]{\frac{x - 5}{2}}$

15. $f^{-1}(x) = -|x|, x \geq 0$

Lesson 7.4 *continued*

16. Restrictions on the domain must be made in inverse functions of all functions where n is even.

17. No. $f^{-1}(x) = \frac{1}{3}x$ and $\frac{1}{f(x)} = \frac{1}{3x} \Rightarrow \frac{1}{3}x \neq \frac{1}{3x}$

$g^{-1}(x) = \frac{3}{2}x$ and $\frac{1}{g(x)} = \frac{3}{2x} \Rightarrow \frac{3}{2}x \neq \frac{3}{2x}$

18.

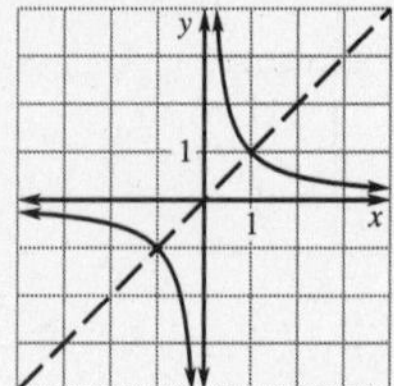

19. $f(f(x)) = f\left(\frac{1}{x}\right) = \frac{1}{\frac{1}{x}} = x$

20. yes; $f(g(x)) = g(f(x)) = x$ **21.** no **22.** yes

23. yes

Reteaching with Practice

1. $y = \frac{1}{4}x - 2$ **2.** $y = -\frac{1}{3}x + 4$

3. $y = \frac{3}{2}x + 6$ **4.** $f(g(x)) = x = g(f(x))$

5. $f(g(x)) = x = g(f(x))$ **6.** $\sqrt{-x}$

7. $-\frac{\sqrt[3]{x}}{3}$ **8.** $\sqrt[4]{x}$ **9–14.** Check graphs.

9. yes; $f^{-1}(x) = \frac{x+4}{3}$ **10.** not a function

11. yes; $f^{-1}(x) = \sqrt[3]{x+4}$ **12.** not a function

13. yes; $f^{-1}(x) = \sqrt[3]{-x}$ **14.** not a function

Real-Life Application

1. $R = \frac{3}{2}(H - 1)$ **2.** 7 ft 6 in. **3.** 8 ft 3 in.

4. $v = \sqrt{672 - 64R}$

5.

Height, H	5.5	6.0	6.5	7.0
Reach, R	6.75	7.5	8.25	9
Velocity, v	15.49	13.86	12	9.80

Challenge: Skills and Applications

1. a. $f^{-1}(x) = \frac{1}{m}(x - b)$ **b.** $\left(\frac{-b}{m-1}, \frac{-b}{m-1}\right)$

c. -40

2. a. (b, a) **b.** They are inverse functions.

c. The slope of the tangent to the graph of $y = f^{-1}(x)$ at $x = f(a)$ is the reciprocal of the slope of the tangent to the graph of $y = f(x)$ at $x = a$.

3. $f^{-1}(x) = \begin{cases} x - 1 \text{ if } x \geq 1 \\ \frac{1}{2}(x - 1) \text{ if } x < 1 \end{cases}$

4. $f^{-1}(x) = \begin{cases} \sqrt{x} \text{ if } x \geq 0 \\ -\sqrt{-x} \text{ if } x < 0 \end{cases}$

5. $f(g(a)) = 0.a_1a_2a_3 \ldots = a$;
$g(f(a)) = 0.0a_2a_3a_4 \ldots \neq a$; no

Quiz 2

1. $5x^2 - 5x^{\frac{1}{3}}$, all real numbers

2. $20x^{\frac{7}{3}} - 4x^{\frac{2}{3}}$, all real numbers

3. $5x^2 + 3x$, all real numbers

4. $\frac{5x^{\frac{5}{3}} - 1}{4}$, all real numbers except 0

5. $\frac{4}{x-6}$; $x \neq 6$ **6.** $\frac{4}{x} - 6$; $x \neq 0$ **7.** x; $x \neq 0$

8. $3[(x + 2)/3] - 2 = x$; $(3x - 2 + 2)/3 = x$

9. $[(x - 2)^{1/4}]^4 + 2 = x$; $(x^4 + 2 - 2)^{1/4} = x$

10. $f^{-1}(x) = \frac{x - 7}{2}$ **11.** $g^{-1}(x) = \left(\frac{x}{3}\right)^{\frac{1}{4}}$

12. no **13.** yes

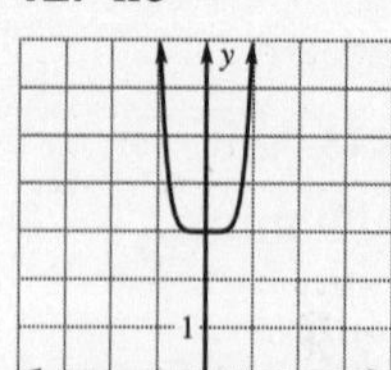

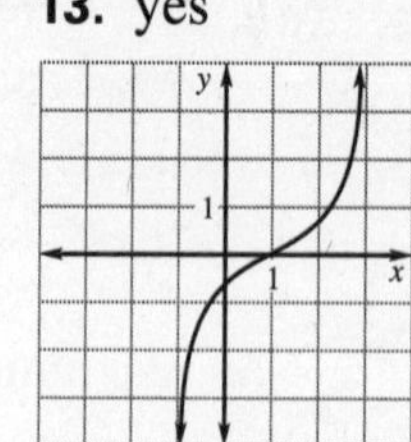

Lesson 7.5

Warm-up Exercises

1.

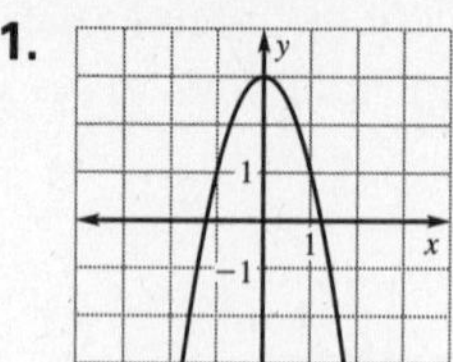

2.

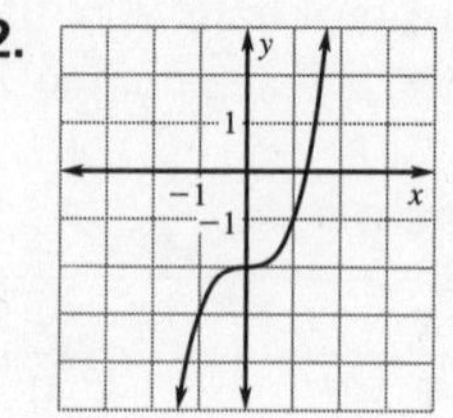

Daily Homework Quiz

1.

x	4	2	0	2	4
y	-2	-1	0	1	2

2. $y = \frac{1}{3}(x + 4)$ **3.** $f^{-1}(x) = \sqrt[5]{\frac{x}{2}}$

4. $f^{-1}(x) = \sqrt[3]{\frac{x + 1}{4}}$

Lesson Opener

Allow 10 minutes.

1.

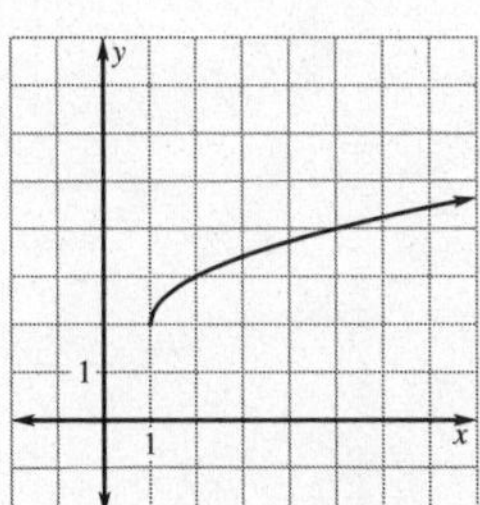

2.

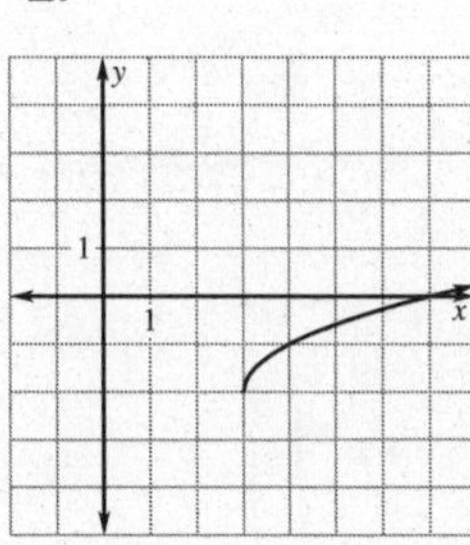

3.

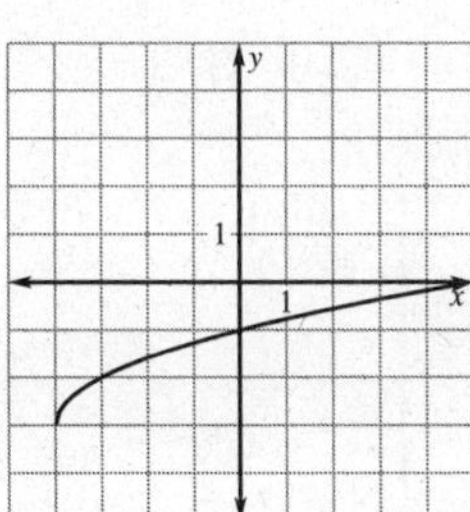

4.

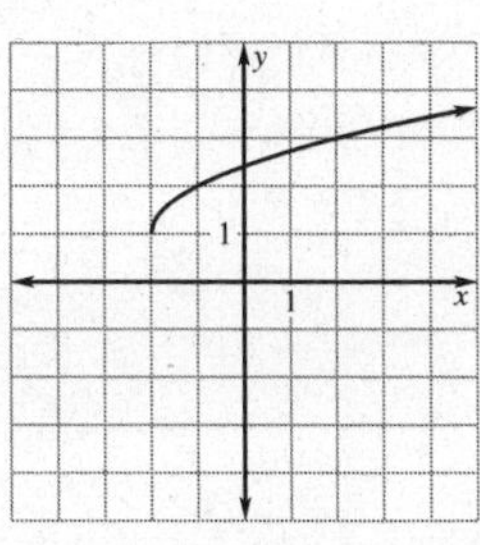

Practice A

1. F **2.** C **3.** B **4.** E **5.** A **6.** D

7. Shift the graph 3 units up. **8.** Shift the graph 2 units down. **9.** Reflect the graph across the x-axis. **10.** Shift the graph 1 unit left. **11.** Shift the graph 4 units right. **12.** Stretch the graph vertically by a factor of 2. **13.** Shift the graph 3 units down. **14.** Shift the graph 2 units up. **15.** Shift the graph 7 units left. **16.** Shift the graph 5 units right. **17.** Shrink the graph vertically by a factor of $\frac{1}{2}$. **18.** Reflect the graph across the x-axis.

19.

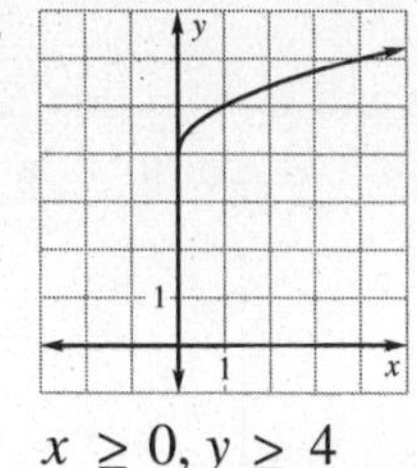

$x \geq 0, y \geq 4$

20.

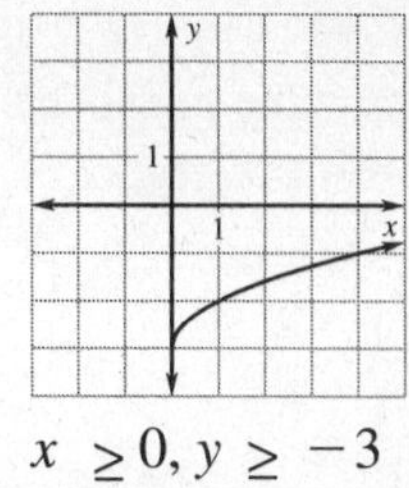

$x \geq 0, y \geq -3$

21.

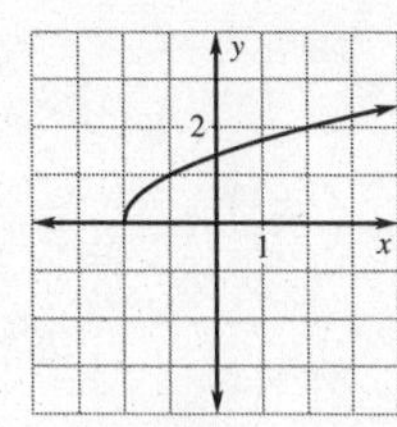

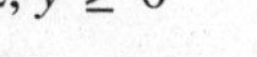

$x \geq -2, y \geq 0$

22.

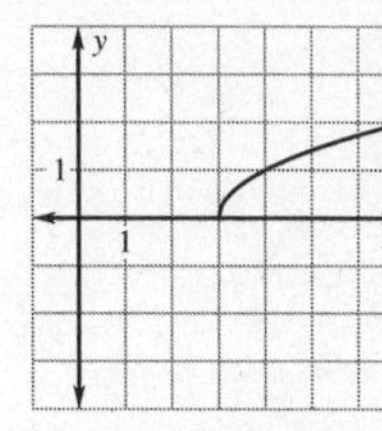

$x \geq 3, y \geq 0$

23.

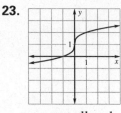

x, y are all real numbers.

24.

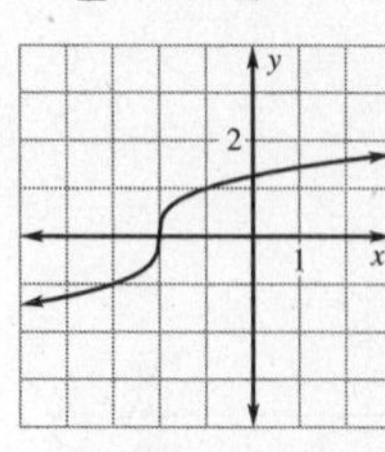

x, y are all real numbers.

25. Domain: $0 \leq h \leq 100$, Range: $0 \leq t \leq 2.5$

26.

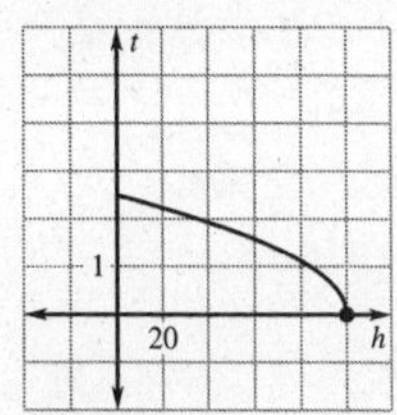

27. 36 ft

Practice B

1. E **2.** B **3.** F **4.** A **5.** C **6.** D

7. Shift the graph 4 units left and 3 units up.

8. Shift the graph 4 units left and 2 units down.

9. Shift the graph 4 units left and reflect it across the x-axis. **10.** Shift the graph 4 units right and 3 units down. **11.** Shift the graph 4 units right and 2 units up. **12.** Shift the graph 4 units right, reflect across the x-axis, and shift 2 units up. **13.** Reflect the graph across the x-axis and shift 1 unit down. **14.** Reflect the graph across the x-axis and shift 1 unit up. **15.** Shift the graph 1 unit right and 5 units up. **16.** Shift the graph 1 unit left and 5 units up.

Lesson 7.5 *continued*

17. Shift the graph 1 unit left and 2 units down.

18. Shift the graph 1 unit right and 2 units down.

19.

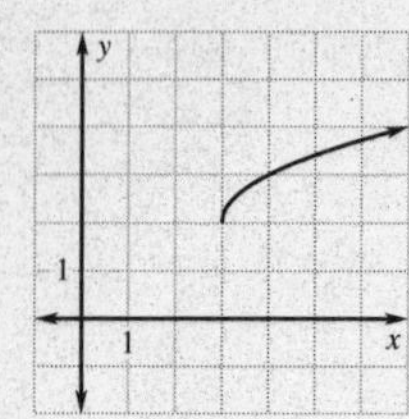

$x \geq 3, y \geq 2$

20.

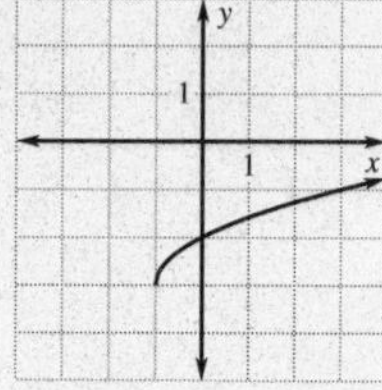

$x \geq -1, y \geq -3$

21.

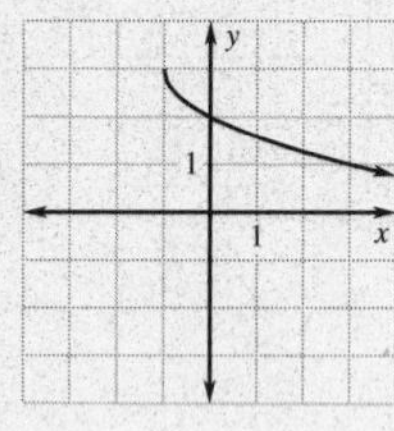

$x \geq -1, y \leq 3$

22.

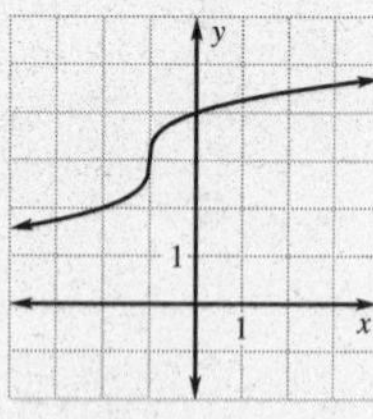

x, y are all real numbers.

23.

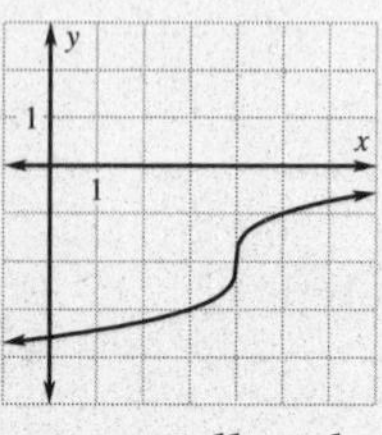

x, y are all real numbers.

24.

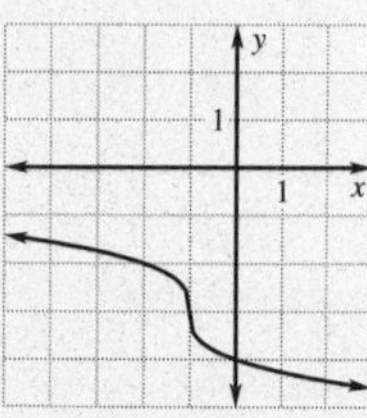

x, y are all real numbers.

25. Domain: $t \geq -273$, Range: $v \geq 0$

26.

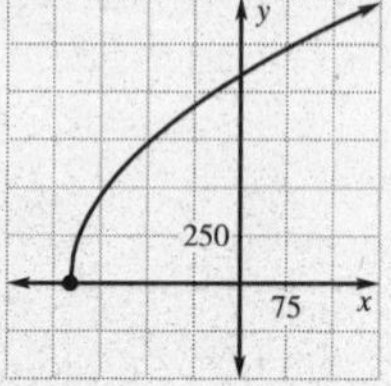

27. 11.58°C

Practice C

1. B **2.** A **3.** C

4.

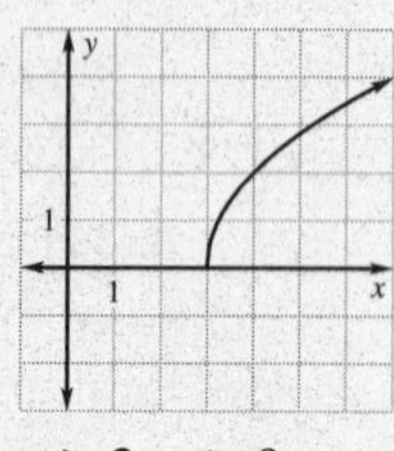

$x \geq 3, y \geq 0$

5.

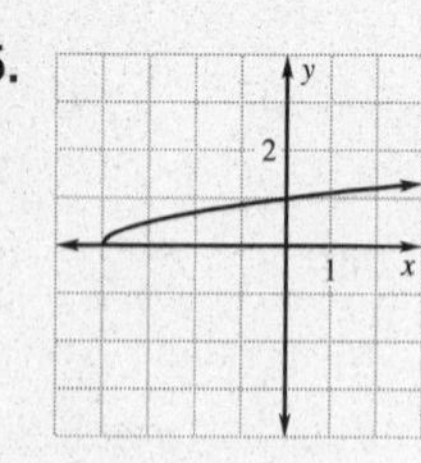

$x \geq -4, y \geq 0$

6.

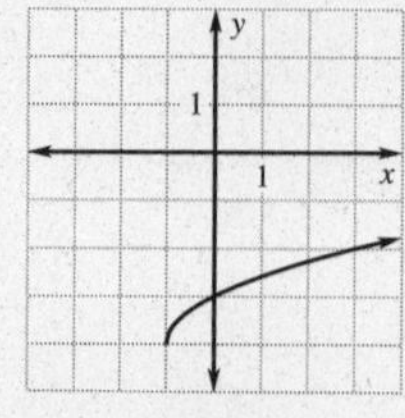

$x \geq -1, y \geq -4$

7.

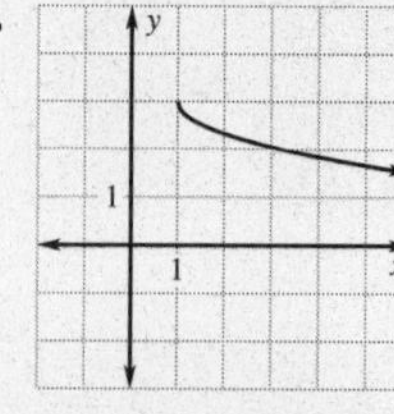

$x \geq 1, y \leq 3$

8.

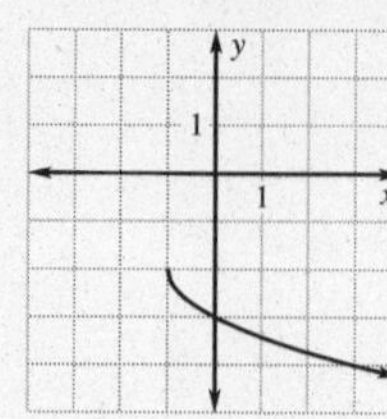

$x \geq -1, y \leq -2$

9.

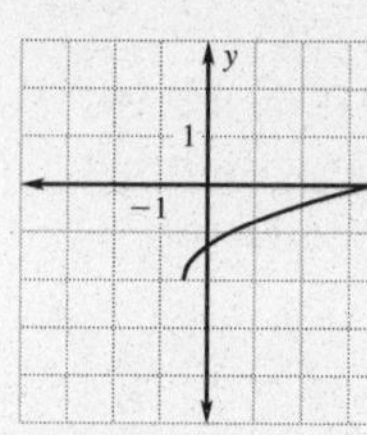

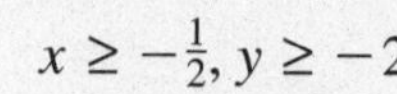

$x \geq -\frac{1}{2}, y \geq -2$

10.

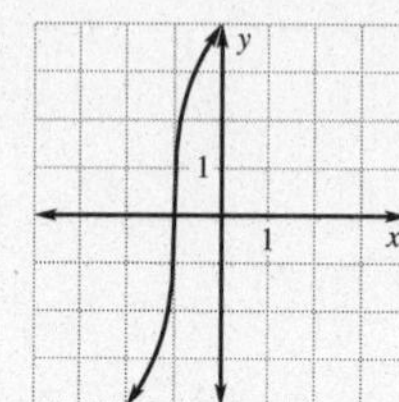

x, y are all real numbers.

11.

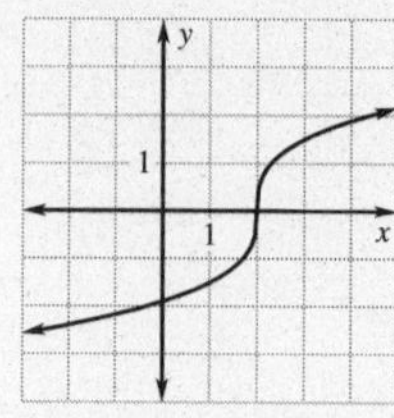

x, y are all real numbers.

12.

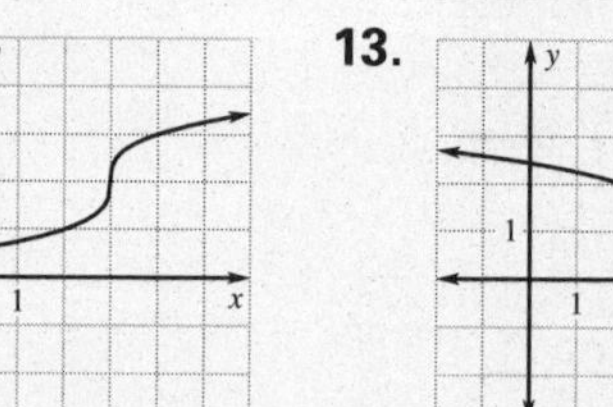

x, y are all real numbers.

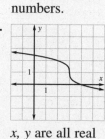

13. x, y are all real numbers.

14.

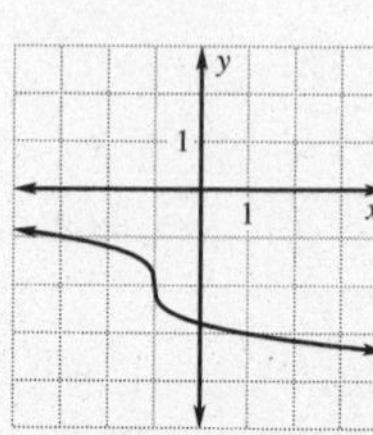

x, y are all real numbers.

15.

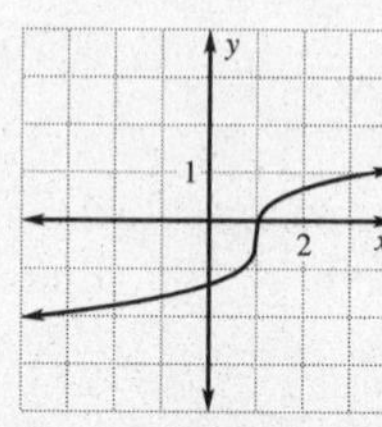

x, y are all real numbers.

16.

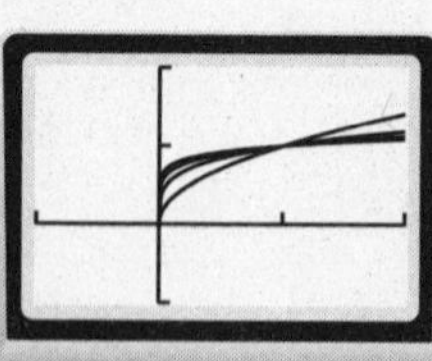

(1, 1) and (0, 0)

17. On the interval (0, 1), the larger the root, the steeper the graph. On the interval $(1, \infty)$, the larger the root the less steep the graph.

18.

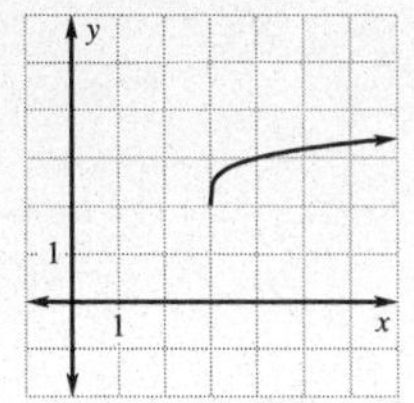

19.

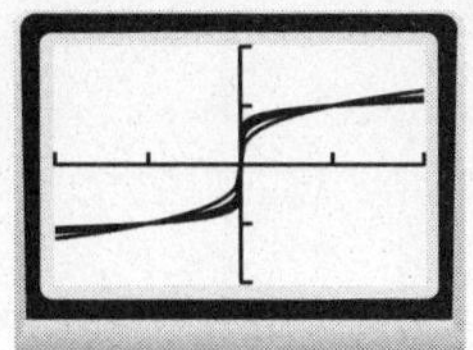

$(-1, -1)$, $(0, 0)$, and $(1, 1)$

20. On the interval $(-1, 1)$, the larger the root, the steeper the graph. On the intervals $(-\infty, -1)$ and $(1, \infty)$, the larger the root, the less steep the graph. **21.**

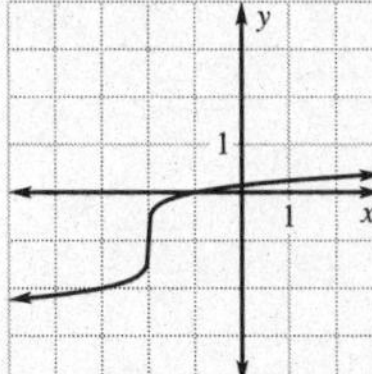

22.

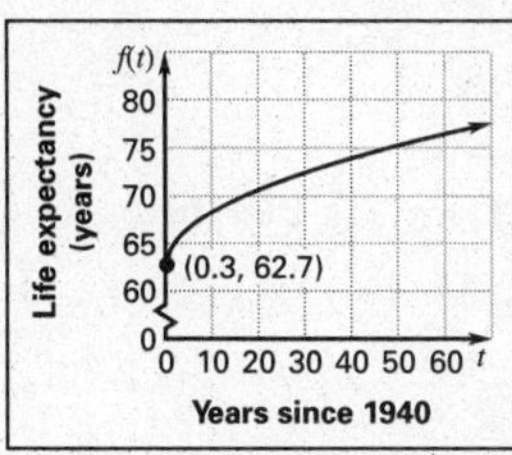

1994

Reteaching with Practice

1. shift up 3 units **2.** shift left 5 units

3. shift right 1 unit and down 4 units

4. shift left 1 unit and up 6 units

5.

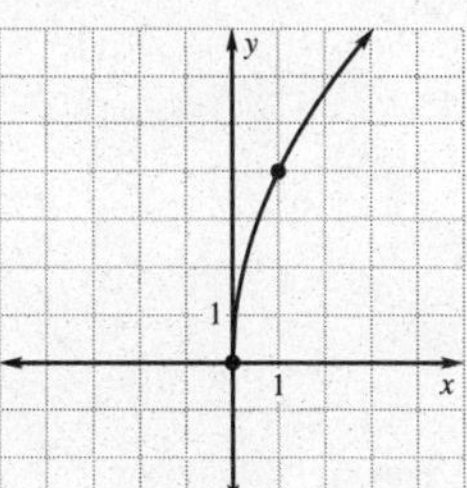

6.

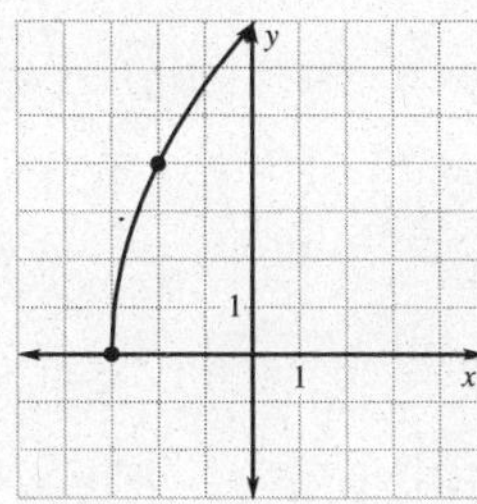

7.

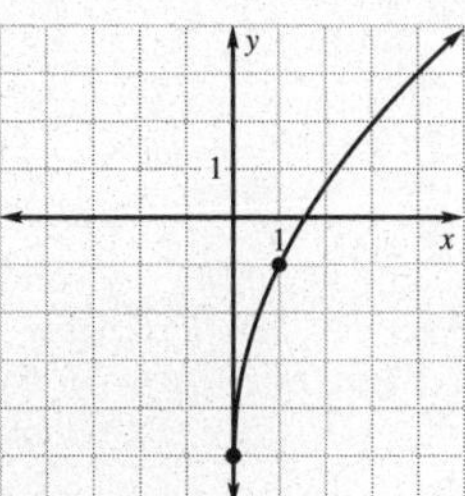

8.

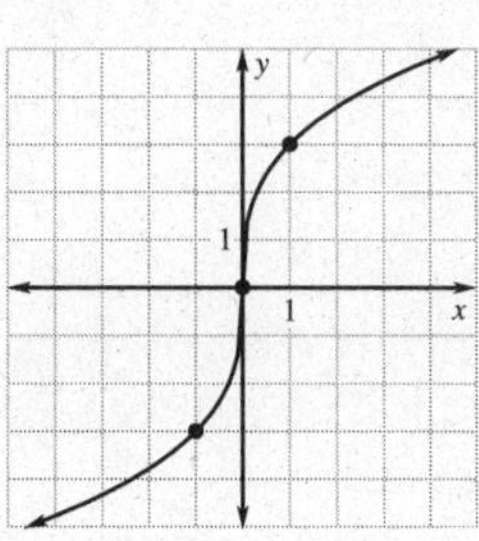

9.

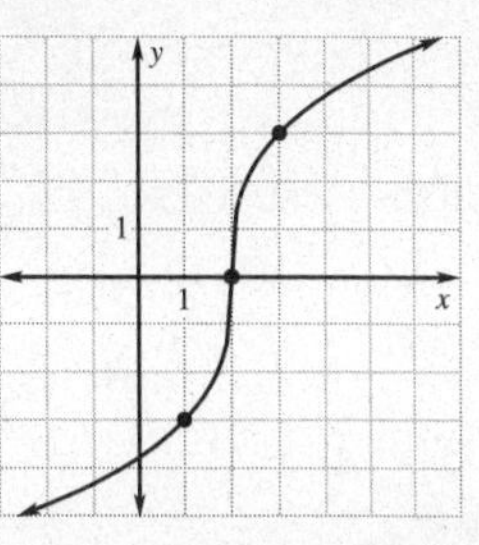

10.

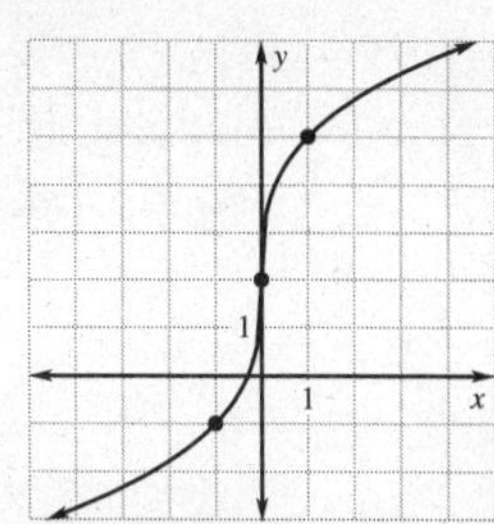

11. $x \geq 0, y \geq 0$ **12.** $x \geq -3, y \geq 0$

13. $x \geq 0, y \geq -5$ **14.** x, y are all real numbers.

15. x, y are all real numbers.

16. x, y are all real numbers.

Interdisciplinary Application

1.

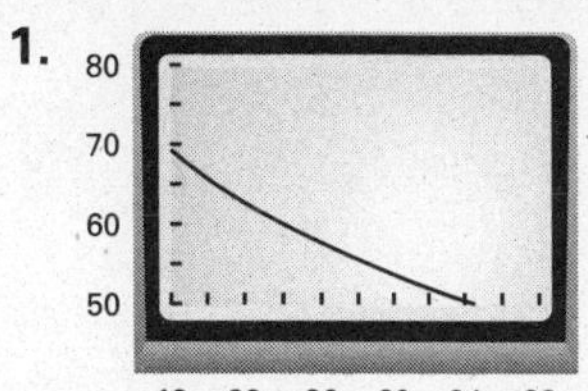

2. 23.6

3. 146.7 lb

4. 152.1 lb and 178.5 lb **5.** 27.6, 23

Challenge: Skills and Applications

1.

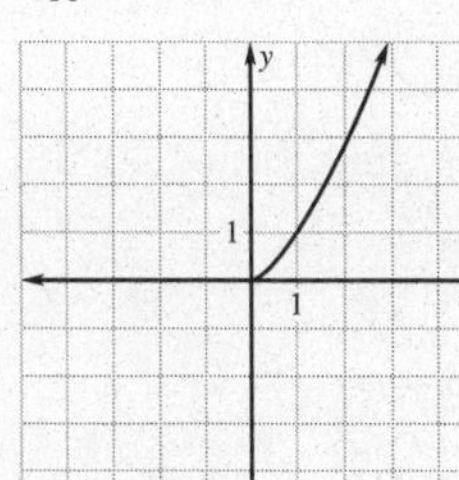

Domain: $x \geq 0$;
Range: $x \geq 0$

2.

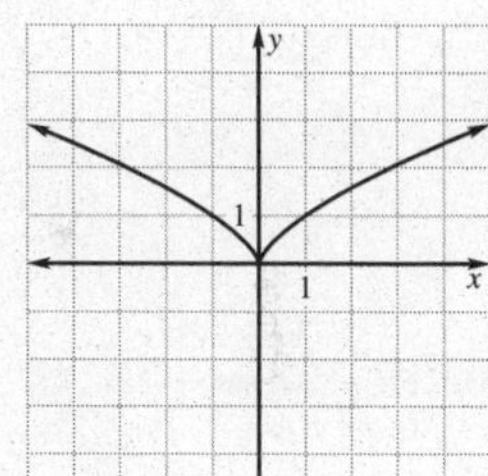

Domain: all real numbers
Range: $x \geq 0$

3. a. $(1, 1)$ **b.** $y = x^{1/3}$ **c.** $y = x^{1/2}$

d.

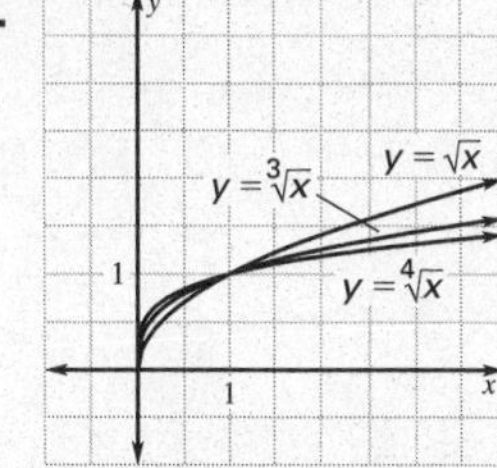

e. The values of $\sqrt[n]{x}$ approach 1 for any $x \neq 0$.

4. $(0, 1)$, $(-1, 0)$, $(8, 3)$ **5.** $(64, 8)$ **6.** $(26, 3)$

Answers

Lesson 7.5 *continued*

7. (0, 2) **8. a.** $(0, 0) + P = -(-P) = P$, for all points P on the graph. **b.** $(-a, -b)$; the line through (a, b) and $(-a, -b)$ also passes through (0, 0). **c.** (8, 2) + (27, 3) = (125, 5)

Lesson 7.6

Warm-up Exercises

1. $x^2 - 6x + 9$ **2.** $(x + 4)^2$ **3.** 6 **4.** 2 **5.** 4

Daily Homework Quiz

1. Shift the graph of f 2 units left and 1 unit down.

2.

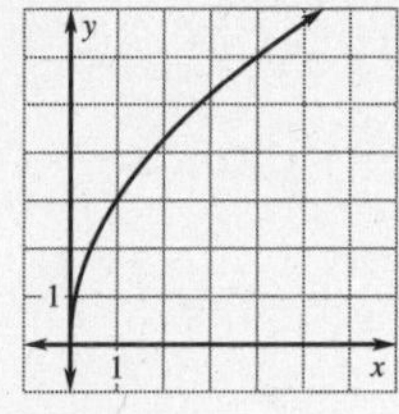

domain: x ‡ 0,
range: y ‡ 0

3.

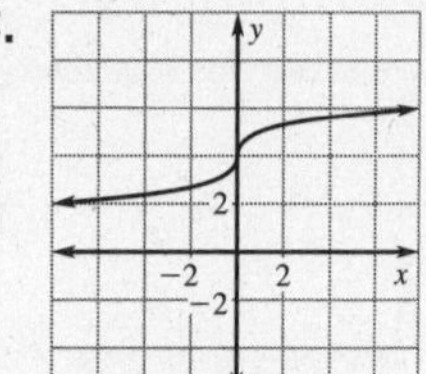

domain, range:
all real numbers

Lesson Opener

Allow 15 minutes.

1. 0.32 **2.** –2, 2 **3.** 0.30 **4.** 2.94 **5.** –10.86, 0.16, 4.70 **6.** –0.83, 2, 4.83 **7.** 1.5 **8.** –2.33, 3.33

Graphing Calculator Activity

1. $x = 32$ **3.** $x = 81$ **5.** $x = 625$

Practice A

1. yes **2.** yes **3.** no **4.** yes **5.** no **6.** yes **7.** 16 **8.** 64 **9.** 64 **10.** 8 **11.** 4 **12.** 125 **13.** $\frac{1}{16}$ **14.** −8 **15.** 81 **16.** 1 **17.** 27 **18.** −27 **19.** 3 **20.** no solution **21.** $\frac{4}{3}$ **22.** 10 **23.** 3 **24.** 2 **25.** 0.81 ft **26.** 0.20 ft **27.** 3.24 ft **28.** 100 ft **29.** 36 ft **30.** 225 ft

Practice B

1. 8 **2.** 16 **3.** 8 **4.** 9 **5.** 8 **6.** −8 **7.** 12 **8.** 1 **9.** $\frac{5}{3}$ **10.** $\frac{1}{3}$ **11.** 3 **12.** no solution **13.** $\frac{7}{2}$ **14.** 12 **15.** −3 **16.** 4 **17.** 0 **18.** $\frac{3}{2}$ **19.** $\frac{7}{2}$ **20.** no solution **21.** 5 **22.** −2, −1 **23.** 6 **24.** 3 **25.** −3, −4 **26.** 8 **27.** 5 **28.** 2.25 **29.** 3.24 **30.** −6.5 **31.** 3.85 **32.** 6.75 **33.** 1.10 **34.** 9.77 ft **35.** 10.78 ft **36.** 34,722.22 ft

Practice C

1. 65 **2.** −516 **3.** $\frac{5}{2}$ **4.** $\frac{17}{3}$ **5.** no solution **6.** 3 **7.** 18 **8.** no solution **9.** $\frac{6751}{4}$ **10.** $-\sqrt{78}, \sqrt{78}$ **11.** $-\sqrt{10}, \sqrt{10}$ **12.** $-2\sqrt{7}, 2\sqrt{7}$ **13.** −6 **14.** no solution **15.** 7, 8 **16.** 3, 5 **17.** 1 **18.** $-3, \sqrt{6}, -\sqrt{6}$ **19.** 5 **20.** 1, 2, −2 **21.** 4 **22.** $\frac{169}{64}$ **23.** no solution **24.** no solution **25.** $\frac{16}{45}$ **26.** 0 **27.** −1, 3 **28.** 4 in. **29.** 24 in.

Reteaching with Practice

1. −8 **2.** 1 **3.** 81 **4.** 81 **5.** 4 **6.** $\frac{1}{16}$ **7.** 32 **8.** 24 **9.** 7 **10.** 4 **11.** 5 **12.** 2

Cooperative Learning Activity

1. about 4 sec; about 4.04 sec **2.** Answers will vary. **3.** no; it will only take about 5.71 sec, or about 1.67 sec longer

Real-Life Application

1. 1.25 in. **2.** 0.5 in.

3. $T = \left(1 + \frac{A}{50{,}000}\right)(136.4\sqrt{p} + 4.5)$

4. 2.05 in.

Math and History Applications

1. 712 km/h **2.** about 1262 meters **3.** 30 minutes **4.** about 4 m/sec **5.** about 92 meters

Challenge: Skills and Applications

1. 5 **2.** −1 **3.** 7 **4.** $3, \frac{9}{4}$ **5.** 2, −1 **6.** 0, −2 **7.** $\frac{49}{8}$ **8.** $\frac{7}{3}$ **9.** 10 **10.** 9 **11.** 4 **12.** 8, 0 **13.** no solution

14. a. $x^2 - 2x + 1 = 4; x^2 - 2x - 3 = 0;$ $x = 3, -1$ **b.** 3 checks; −1 does not

c. When both sides were squared, an extraneous root was introduced, because for any real number a, $(-a)^2 = a^2$.

Lesson 7.7

Warm-up Exercises

1. 10.5 **2.** 20.25 **3.** 8.5 **4.** 5.23 **5.** 3.35

Daily Homework Quiz

1. 64 **2.** 27 **3.** 20

Lesson Opener

Allow 15 minutes.

1. 60.24 **2.** 342.43 **3.** 60.24 **4.** 56.8 **5.** 30.42 **6.** 48 **7.** *Sample answer:* The sets of data are similar because the means are about the same; they are different because the values in the second set are more spread out. (*Note:* This question is intended to foreshadow the topic of standard deviation; the data set in problem 3 has a higher standard deviation than the data set in problem 1.)

Practice A

1. 1, 1, 1, 2, 2, 3, 4, 5, 5, 6, 7, 7, 8, 8, 9; 4.6; 5; 1 **2.** 8, 10, 10, 10, 12, 12, 13, 15, 16, 19; 12.5; 12; 10 **3.** 24 **4.** 48 **5.** 3 **6.** 7 **7.** 149 **8.** 21 **9.** 7.5, 14 **10.** 136, 154.5 **11.** 1, 4.5 **12.** 35, 42

13.

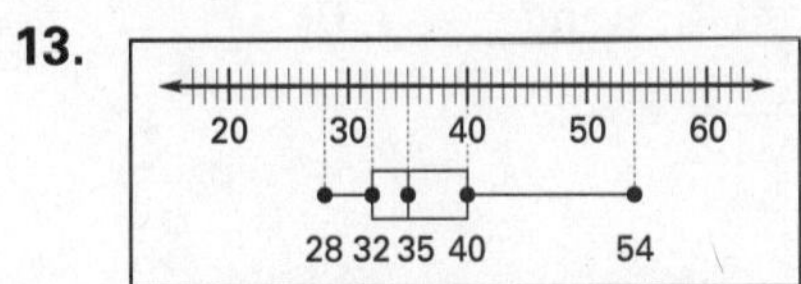

14.

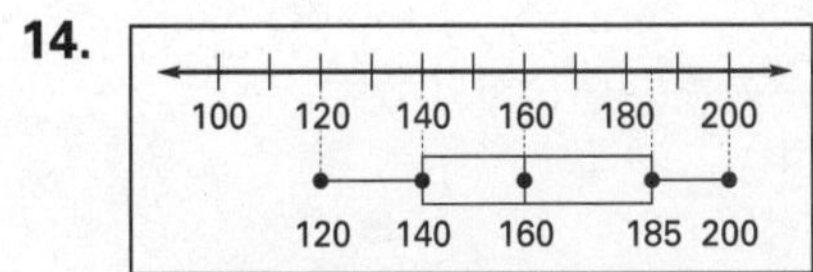

15.

Interval	*Tally*	*Frequency*
1–2	𝍸 ||	7
3–4	||||	4
5–6	||||	4
7–8	𝍸 ||	7
9–10	|||	3

16.

Interval	*Tally*	*Frequency*
1–2	𝍸 |||	8
3–4	|||	3
5–6	|||	3
7–8		0
9–10	|||	3

17. Exercise 16 **18.** Exercise 15

Practice B

1. 11.4; 9; 8 **2.** 20.4; 22; 22 **3.** $49.\overline{5}$; 48; 44 **4.** $127.\overline{2}$;130; 100 **5.** 47, 16.7 **6.** 15.5, 5.01 **7.** ≈ 4.5 **8.** 25.2 oz **9.** 25 oz **10.** 28 oz **11.** 6 **12.** 5; 7 **13.**

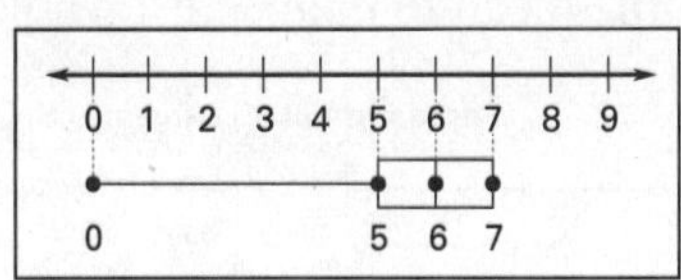

Practice C

1. 3; 1.05 **2.** 0.6; 0.208 **3.** 3; 0.957 **4.** 20; 6.54 **5.** b **6. a.** 6.7, 2 **b.** 23.3, 25.5; The median is a more accurate measure of central tendency when a small number of data is much different than the majority of the data.

7. 30 **8.** 107.4 **9.** 8.052

10. Machine #1: 1.0008; Machine #2: 0.9993

11. Machine #1: 0.00098; Machine #2: 0.0009

12. Machine #2

13.

Children of U.S. Presidents		
Interval	*Tally*	*Frequency*
0–2	𝍸 𝍸 𝍸 |	16
3–5	𝍸 𝍸 𝍸 |	16
6–8	𝍸 ||	7
9–11	|	1
12–14	|	1

Lesson 7.7 *continued*

14.

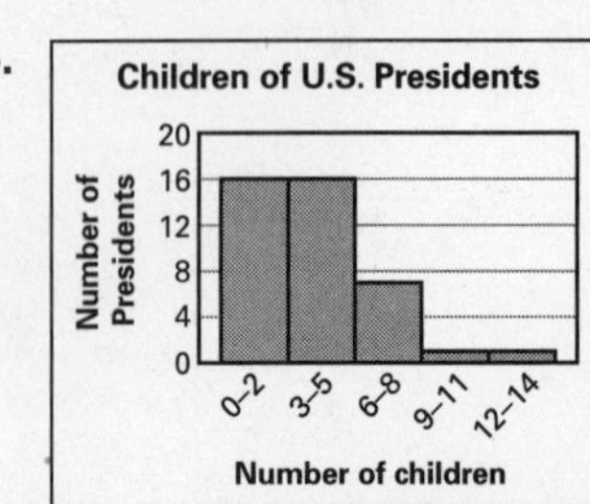

15. median

Reteaching with Practice

1. 14.4, 14.5, 15 **2.** 76.7, 78, no mode
3. 8, 2.19 **4.** 45, 14.6

Interdisciplinary Application

1. 1035; 597; 441, 498 **2.** 6958 **3.** 1267.7 (round the mean to the nearest whole number)

4.

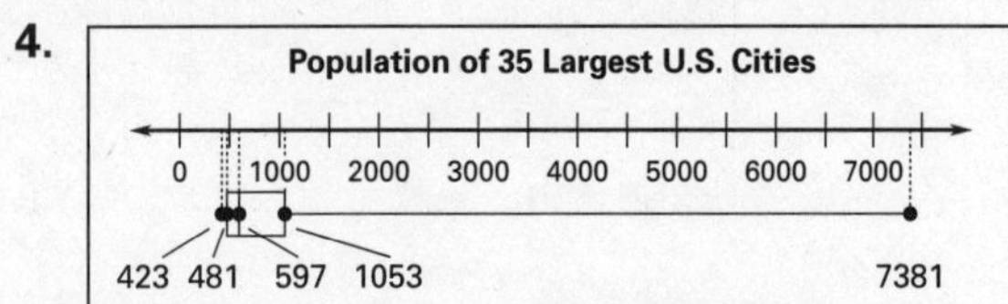

Challenge: Skills and Applications

1. a. −1.52, −0.52, 0.08, 0.48, 1.48 **b.** 0, 1, yes **2. a.** 6, 12, 20; 16, 28, 37
b. $y = \frac{3}{2}x + 7$; (12, 25) **c.** $y = \frac{3}{2}x + 8$

3. a. $\frac{(x_1 - \bar{x})^2 + (x_2 - \bar{x})^2 + (x_3 - \bar{x})^2}{3} =$

$\frac{x_1{}^2 - 2\bar{x}x_1 + (\bar{x})^2 + x_2{}^2 - 2\bar{x}x_2 + (\bar{x})^2 + x_3{}^2 - 2\bar{x}x_3 + (\bar{x})^2}{3} =$

$\frac{1}{3}[x_1{}^2 + x_2{}^2 + x_3{}^2 - 2\bar{x}(x_1 + x_2 + x_3) + 3(\bar{x})^2]$

b. the right side

$= \frac{x_1{}^2 + x_2{}^2 + x_3{}^2}{3} - 2\bar{x}\left(\frac{x_1 + x_2x_3}{3}\right) + (\bar{x})^2$

$= \overline{x^2} - 2(\bar{x})^2 + (\bar{x})^2 = \overline{x^2} = (\bar{x})^2$

Review and Assessment

Review Games and Activities

1. $\frac{1}{27}$ **2.** 9 **3.** $3x - 11$ **4.** $3x - 11$

5. $\pm\sqrt{\frac{x-5}{2}}$ **6.** 9 **7.** 6 **8.** 4 **9.** 4

10. 1.76 **11.** 4 **12.** 6 **13.** 8

With A N I N T R O - D U C T I O N example

A	N	I	N	T	R	O-
1	8	2	9	3	5	7

D	U	C	T	I	O	N
10	13	↓ example	4	6	12	11

Test A

1. −2 **2.** 5 **3.** 9 **4.** $\frac{1}{2}$ **5.** 6 **6.** $2xy^2z$

7. x^6y^6 **8.** $7\sqrt{2}$ **9.** $4x - 5$; Domain: all real numbers **10.** $2x + 5$; Domain: all real numbers

11. $3x^2 - 15x$; Domain: all real numbers

12. $\frac{3x}{x-5}$; Domain: all real numbers except 5

13. $3x - 15$; Domain: all real numbers

14. $f(x) = x - 9$ **15.** $f(x) = 2x - 4$

16. $f(x) = \frac{1}{3}x - 2$

17.

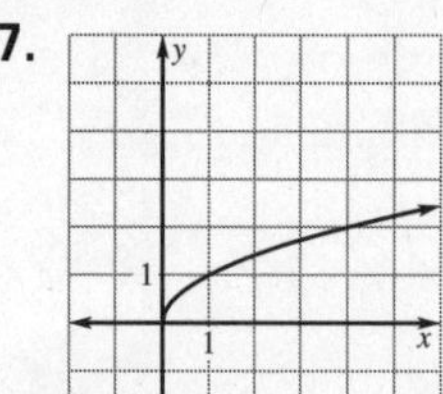

Domain: x ‡ 0
Range: y ‡ 0

18.

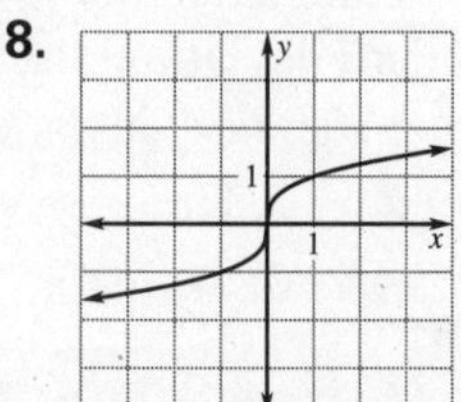

Domain: all real numbers
Range: all real numbers

19.

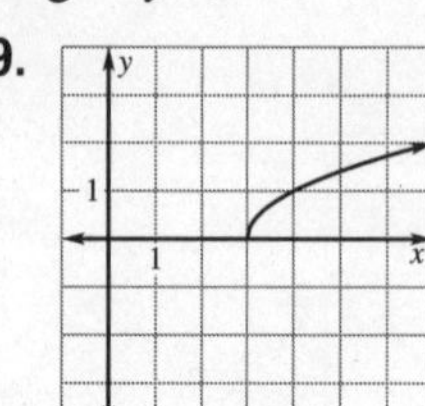

Domain: x ‡ 3
Range: y ‡ 0

20. 1 **21.** 6 **22.** $\pm 3\sqrt{2}$

23. mean = 83.4
median = 85
mode = 84
range = 30
standard deviation = 9.25

24.

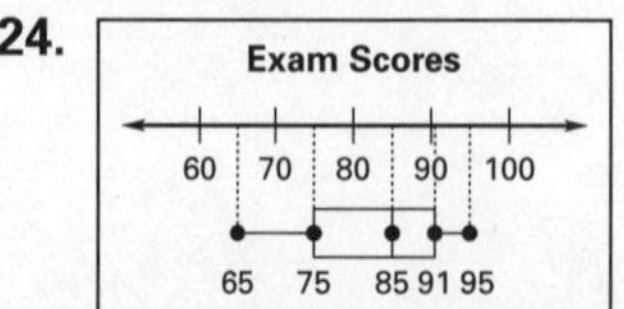

25.

Interval	*Tally*	*Frequency*
60–69	|	1
70–79	| |	2
80–89	| | |	3
90–99	| | | |	4

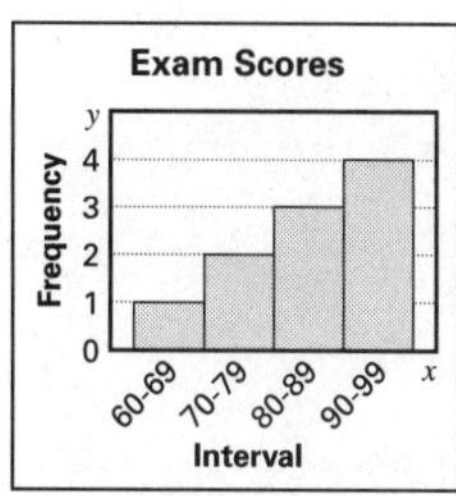

Test B

1. -4 **2.** 9 **3.** 9 **4.** $\frac{1}{5}$ **5.** 12 **6.** $-2xyz$

7. $\frac{1}{4y^3}$ **8.** $8\sqrt{2}$

9. $3x - 1$; Domain: all real numbers

10. $-x - 1$; Domain: all real numbers

11. $2x^2 - 2x$; Domain: all real numbers

12. $\frac{x-1}{2x}$; Domain: all real numbers except 0

13. $2x - 1$; Domain: all real numbers

14. $f(x) = \frac{1}{2}x + 2$ **15.** $f(x) = -\frac{3}{2}x + 6$

16. $f(x) = x^{1/2}$

17. 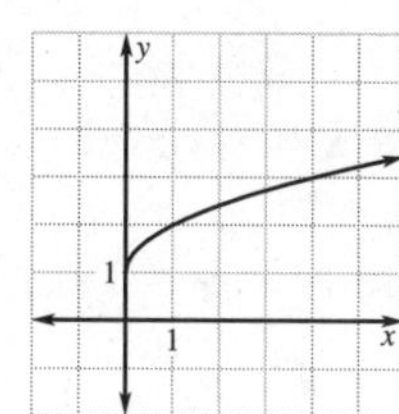

Domain: $x \geq 0$
Range: $y \geq 1$

18.  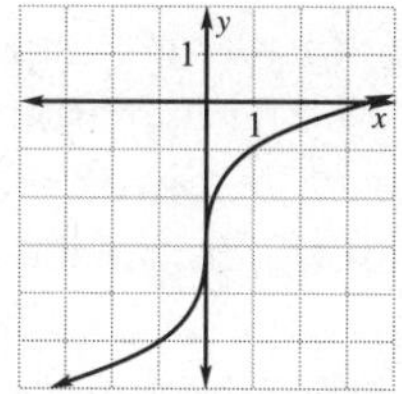

Domain: all real numbers
Range: all real numbers

19. 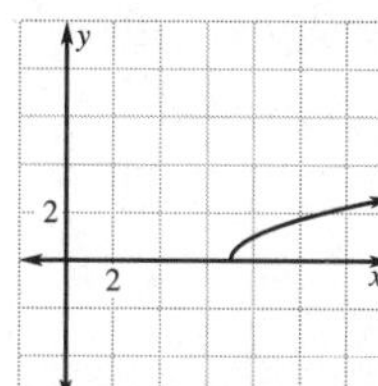

Domain: $x \geq 7$
Range: $y \geq 0$

20. 32 **21.** 3 **22.** $-4, 3$

23. mean = 976.4
median = 971
mode = 964
range = 184
standard deviation = 51.3

24. 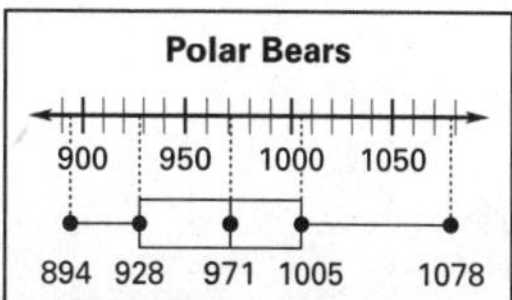

25.

Interval	*Tally*	*Frequency*
875–929	| | |	3
930–984	| | |	3
985–1039	| | |	3
1040–1094	|	1

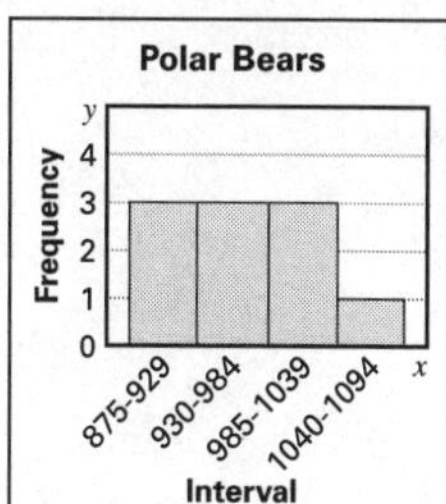

Test C

1. -5 **2.** 9 **3.** 3 **4.** 6 **5.** $3^{5/6}$

6. $2xy\sqrt[4]{2x}$ **7.** $\frac{9x^4}{4y^8}$ **8.** $4\sqrt[3]{2}$

9. $2x$; Domain: all real numbers

10. -2; Domain: all real numbers

11. $x^2 - 1$; Domain: all real numbers

12. $\frac{x-1}{x+1}$; Domain: all real numbers except -1

13. x; Domain: all real numbers

14. $f(x) = \frac{1}{2}x + 2$ **15.** $f(x) = \sqrt{x-5}$; $x \geq 5$

16. $f(x) = x^3 + 7$

17. 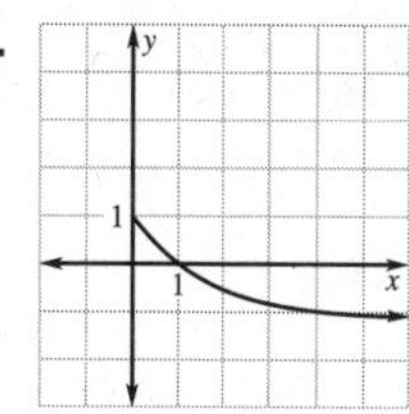

Domain: $x \geq 0$
Range: $y \leq 1$

18. 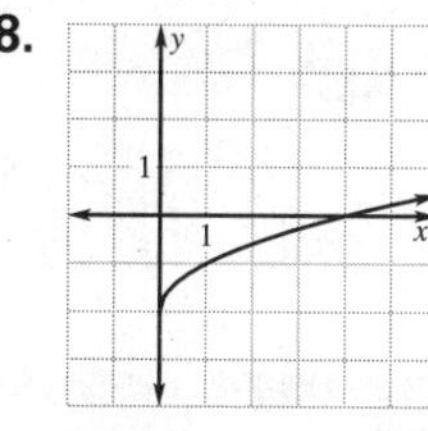

Domain: $x \geq 0$
Range: $y \geq -2$

Review and Assessment *continued*

19. 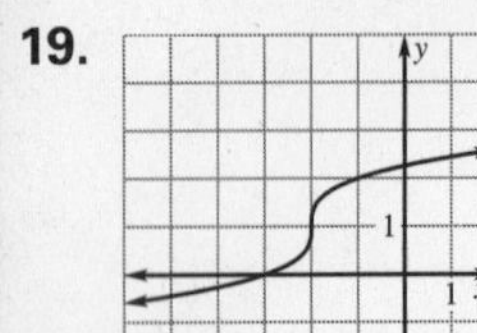

Domain: all real numbers
Range: all real numbers

20. 2, 3 **21.** no solution **22.** 25

23.

	Boys	*Girls*
mean	70.1	65.3
median	70	65
mode	70	65
range	42	23
standard deviation	11.7	6.87

24. girls team

25.

Boys Basketball Points

50 60 70 80 90

48 65 70 80 90

26.

Interval	*Tally*	*Frequency*
55–59	| |	2
60–64	| |	2
65–69	| | |	3
70–74	| |	2
75–79	|	1

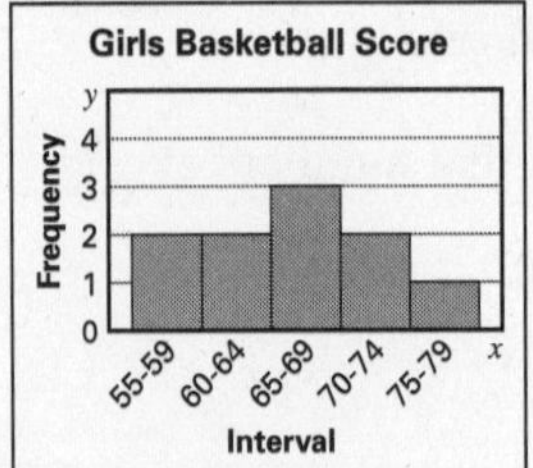

SAT/ACT

1. D **2.** B **3.** B **4.** C **5.** A **6.** D **7.** A **8.** C **9.** D **10.** C

Alternative Assessment

1. a–f. Complete answers should include the following: **a.** • Correct answer is $(27)^{1/3} = 3$. • Explain that Jino divided 27 by 3 rather than finding the cube root of 27.

b. • Jino's answer is correct. **c.** • Correct answer is $\sqrt{x^{16}} = x^8$ • Explain that Jino took the square root of 16 instead of multiplying the exponents $16 \cdot \frac{1}{2}$. **d.** • Correct answer is $\sqrt{81} = 9$. • Explain that Jino should take the square root of the value only once. **e.** • Correct answer is $(4 - \sqrt{x})^2 = 16 - 8\sqrt{x} + x$. • Explain that Jino multiplied incorrectly, forgetting the middle terms. **f.** • Correct answer is $x = -2$ or $x = -1$. • Explain that Jino did not isolate the radical expression before he squared both sides.

2. a. $f^{-1}(x) = x + 3$ **b.** $g^{-1}(x) = \frac{x}{2}$
c. $f(g(x)) = 2x - 3$ **d.** $g(f(x)) = 2x - 6$
e. $[f(g(x))]^{-1} = \frac{x + 3}{2}$
f. $[g(f(x))]^{-1} = \frac{x + 6}{2} = \frac{x}{2} + 3$

3. ***Critical Thinking*** It seems that $[f(g(x))]^{-1} = g^{-1}(f^{-1}(x))$ and $[g(f(x))]^{-1} = f^{-1}(g^{-1}(x))$. This is logical because an inverse function performs the reverse operation as the original function. Therefore, the inverse of a composition of two functions performs the reverse operations in the reverse order.

Project: Walk This Way

1. Check student's measurement. **2.** Check that the set of data is reasonable. **3.** Observe students. **4.** Check that the set of data is reasonable. **5.** Check calculations. **6.** Check the scatter plot. **7.** Check calculations and graph. **8.** Check that the answer is reasonable based on student's previous answers.

Cumulative Review

1. inverse property of addition
2. commutative property of multiplication
3. associative property of multiplication
4. identity property of multiplication
5. distributive property
6. commutative property of addition
7. 4 **8.** 5 **9.** 5, −11 **10.** −1, 6 **11.** 6
12. $\frac{1}{4}$ **13.** C **14.** B **15.** A
16. positive correlation **17.** negative correlation

Review and Assessment *continued*

18. minimum: 0, maximum: 12 **19.** minimum: 1, maximum: 21 **20.** minimum: -5, maximum: 19

21. $\begin{bmatrix} 1 & 1 \\ 2 & 1 \end{bmatrix}$ **22.** $\begin{bmatrix} 4 & 2 \\ 7 & 2 \end{bmatrix}$

23. $\begin{bmatrix} -8 & -4 & -12 \\ -4 & 8 & 8 \end{bmatrix}$ **24.** $\begin{bmatrix} -1 & 6 \\ -3 & 7 \\ -4 & 8 \end{bmatrix}$

25. not defined; number of columns of first matrix is not equal to number of rows of second matrix **26.** $[-1]$ **27.** 8 **28.** 19 **29.** -70

30. $\begin{bmatrix} 1 & 2 \\ \frac{3}{2} & \frac{7}{2} \end{bmatrix}$ **31.** $\begin{bmatrix} \frac{1}{10} & \frac{1}{10} \\ \frac{1}{5} & -\frac{4}{5} \end{bmatrix}$ **32.** $\begin{bmatrix} \frac{1}{6} & -\frac{1}{12} \\ \frac{1}{3} & -\frac{2}{3} \end{bmatrix}$

33. $(2x^2 + 1)^2$ **34.** $(a + 2)^2(a - 2)^2$

35. $(x^2 + y^2)(x + y)(x - y)$

36. $(2a - 5)(4a^2 + 10a + 25)$

37. $(x^2 + y^2)(x - y)$

38. $(3x - 2y)(9x^2 + 6xy + 4y^2)$

39.

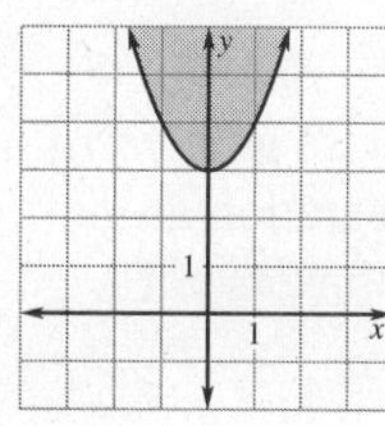

40.

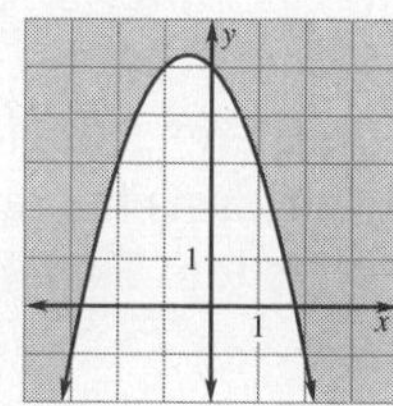

41.

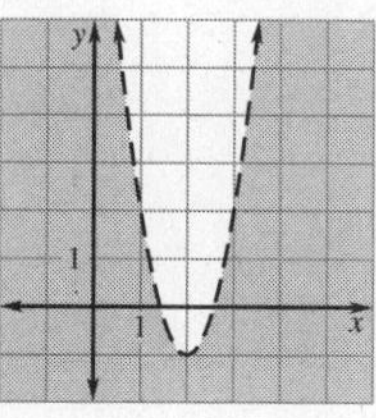

42.

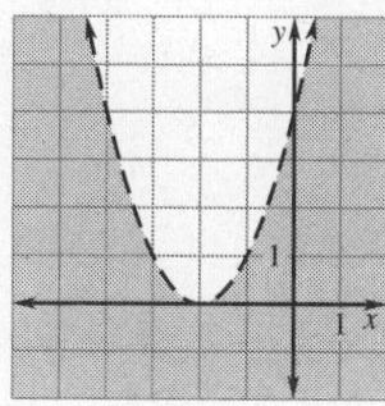

43.

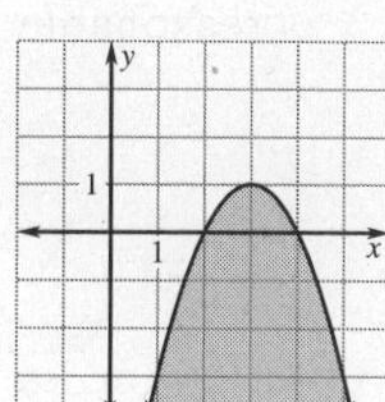

44.

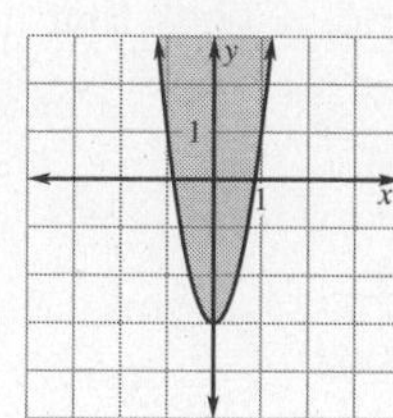

45. $y = 3(x - 3)^2 - 1$ **46.** $y = (x + 2)^2$

47. $y = -2(x + 4)(x - 2)$

48. $y = -(x - 2)(x - 4)$

49. $y = \frac{2}{3}x^2 + 2x - 1$

50. x^3 **51.** $\frac{1}{x^9y^{12}}$ **52.** x^{10} **53.** $\frac{1}{9x^4y^6}$

54. x^3y^5 **55.** $\frac{-5}{x^4}$ **56.** 1254°F

57. Drivers $\geq$ 35 years